MILITARY NANOTECHNOLOGY

This book is the first systematic and comprehensive presentation of the potential military applications of nanotechnology (NT). After a thorough introduction and overview of nanotechnology and its history, it presents the actual military NT R&D in the USA and gives a systematic description of the potential military applications of NT that within 10–20 years may include extremely small computers, miniature sensors, lighter and stronger materials in vehicles and weapons, autonomous systems of many sizes and implants in soldiers' bodies. These potential applications are assessed from a viewpoint of international security, considering the new criteria of dangers for arms control and the international law of warfare, dangers for stability through potential new arms races and proliferation, and dangers for humans and society.

Although some applications (e.g. sensors for biological-warfare agents) could contribute to better protection against terrorist attacks or to better verification of compliance with arms-control treaties, several potential uses, like metal-free firearms, small missiles, or implants and other body manipulation raise strong concerns. For preventive limitation of these potentially dangerous applications of NT, specific approaches are proposed that balance positive civilian uses and take into account verification of compliance.

This book will be of much interest to students of strategic studies, peace studies, conflict resolution and international security, as well as specialists in the fields of military technology and chemical-biological weapons.

Jürgen Altmann holds a PhD in physics and is a recognized expert in the field of disarmament and arms control. He has been working on disarmament-related issues since 1985 conducting research on co-operative verification of disarmament or peace-keeping agreements using automatic sensor systems and assessment and preventive limitations of new military technologies, with major studies on acoustic weapons and micro-systems technologies.

CONTEMPORARY SECURITY STUDIES

NATO'S SECRET ARMY
Operation Gladio and terrorism in Western Europe
Daniel Ganser

THE US, NATO AND MILITARY BURDEN-SHARING
Peter Kent Forster and Stephen J. Cimbala

RUSSIAN GOVERNANCE IN THE TWENTY-FIRST CENTURY
Geo-strategy, geopolitics and new governance
Irina Isakova

THE FOREIGN OFFICE AND FINLAND 1938–1940
Diplomatic sideshow
Craig Gerrard

RETHINKING THE NATURE OF WAR
Isabelle Duyvesteyn and Jan Angstrom (eds)

PERCEPTION AND REALITY IN THE MODERN YUGOSLAV CONFLICT
Myth, falsehood and deceit 1991–1995
Brendan O'Shea

THE POLITICAL ECONOMY OF PEACEBUILDING IN POST-DAYTON BOSNIA
Tim Donais

THE DISTRACTED EAGLE
The rift between America and Old Europe
Peter H. Merkl

THE IRAQ WAR
European perspectives on politics, strategy, and operations
Jan Hallenberg and Håkan Karlsson (eds)

STRATEGIC CONTEST
Weapons proliferation and war in the Greater Middle East
Richard L. Russell

PROPAGANDA, THE PRESS AND CONFLICT
The Gulf War and Kosovo
David R. Willcox

MISSILE DEFENCE
International, regional and national implications
Bertel Heurlin and Sten Rynning (eds)

GLOBALISING JUSTICE FOR MASS ATROCITIES
A revolution in accountability
Chandra Lekha Sriram

ETHNIC CONFLICT AND TERRORISM
The origins and dynamics of civil wars
Joseph L. Soeters

GLOBALISATION AND THE FUTURE OF TERRORISM
Patterns and predictions
Brynjar Lia

NUCLEAR WEAPONS AND STRATEGY
The evolution of American nuclear policy
Stephen J. Cimbala

NASSER AND THE MISSILE AGE IN THE MIDDLE EAST
Owen L. Sirrs

WAR AS RISK MANAGEMENT
Strategy and conflict in an age of globalised risks
Yee-Kuang Heng

MILITARY NANOTECHNOLOGY
Potential applications and preventive arms control
Jürgen Altmann

MILITARY NANOTECHNOLOGY

Potential applications and preventive arms control

Jürgen Altmann

LONDON AND NEW YORK

First published 2006
by Routledge
2 Park Square, Milton Park, Abingdon, Oxon OX14 4RN

Simultaneously published in the USA and Canada
by Routledge
270 Madison Ave, New York, NY 10016

Routledge is an imprint of the Taylor & Francis Group

Transferred to Digital Printing 2006

Typeset in Times by Wearset Ltd, Boldon, Tyne and Wear

British Library Cataloguing in Publication Data
A catalogue record for this book is available from the British Library

Library of Congress Cataloging in Publication Data
A catalog record for this book has been requested

ISBN10: 0-415-37102-3 (hbk)
ISBN10: 0-415-40799-0 (pbk)

ISBN13: 978-0-415-37102-5 (hbk)
ISBN13: 978-0-415-40799-1 (pbk)

Printed and bound by CPI Antony Rowe, Eastbourne

CONTENTS

FIGURES

TABLES

PREFACE

This work was done at Lehrstuhl Experimentelle Physik III, Universität Dortmund. I want to thank Prof. Dieter Suter for making the project application and for general support, including critical reading of draft manuscripts.

The project was funded from February 2002 to April 2003 by the German Foundation for Peace Research (DSF) in its first round of research support. I should like to thank DSF for the funding and the DSF staff at Osnabrück for efficient and co-operative administration.

The work was carried out in the context of the joint projects on preventive arms control (Projektverbund Präventive Rüstungskontrolle PRK) of the Research Association Science, Disarmament and International Security FONAS.

Interviews (by phone or visit) were done with scientists from several research and development institutions, among them Fraunhofer-Institut für Biomedizinische Technik (IBMT), St. Ingbert, Germany, and CeNTech, Münster, Germany. Thanks go to all these institutions and persons.

I want to thank the Bonn International Center for Conversion (BICC) where I was employed from August to December 2003 to work on the convergence of nano-, bio-, information and cognitive science and technology; this half-time employment provided opportunities to continue working on this book.

Thanks go to Mark A. Gubrud for helpful remarks on parts of the text. I am grateful to Götz Neuneck for comments on the sections on preventive arms control and to Kathryn Nixdorff and Jan van Aken for comments on the parts on chemical and biological weapons.

For giving permission to reprint their figures, I should like to thank the NASA Ames Research Center (M. Meyyappan) and the MIT Institute for Soldier Nanotechnologies (E. Downing).

Finally, I want to thank Steve Turrington for his efficient editing.

Dortmund, April 2004
Jürgen Altmann

ABBREVIATIONS

ABM	Antiballistic Missile (System)
AFM	Atomic force microscope
AFOSR	Air Force Office of Scientific Research (USA)
AFRL	Air Force Research Laboratory (USA)
AI	Artificial intelligence
ARO	Army Research Office (USA)
ARL	Army Research Laboratory (USA)
ASIC	Application-specific integrated circuit
ATP	Adenosine triphosphate
AVT	Applied Vehicle Technology Panel of NATO RTO
BICC	Bonn International Center for Conversion
BMBF	Bundesministerium für Bildung und Forschung (Federal Ministry of Education and Research, Germany)
BMDO	Ballistic Missile Defense Organization (USA)
B(T)WC	Biological (and Toxin) Weapons Convention
CBRE	Chemical-Biological-Radiological-Explosive
CBW	Chemical/biological warfare
CEA	Commissariat à l'Energie Atomique (France)
CFE	Conventional Armed Forces in Europe (Treaty)
CMOS	Complementary metal-oxide semiconductor
CNID	Center for Nanoscale Innovation for Defense (Univ. of California, USA)
CNT	Carbon nanotube
CSBM	Confidence and security building measures
CTBTO	Comprehensive Test Ban Treaty Organization
CWC	Chemical Weapons Convention
DARPA	Defense Advanced Research Projects Agency (USA)
DDT	Dichlorodiphenyltrichloroethane
DERA	Defence Evaluation and Research Agency (UK, formerly)

DFG	Deutsche Forschungsgemeinschaft (German Research Foundation)
DGA	Délégation Générale pour l'Armement (France)
DNA	Deoxyribonucleic acid
DoC	Department of Commerce (USA)
DoD	Department of Defense (USA)
DoE	Department of Energy (USA)
DRAM	Dynamic random-access memory
DSF	Deutsche Stiftung Friedensforschung (German Foundation for Peace Research)
DSTO	Defence Science and Technology Organisation (Australia)
DUSD (R)	Deputy Undersecretary of Defense for Research (USA)
EDIG	European Defence Industry Group
EM	Electromagnetism/ic
EUCLID	European Cooperation for the Long term In Defence
FOI	Totalförsvarets Forskningsinstitut (Swedish Defence Research Agency)
FONAS	Forschungsverbund Naturwissenschaft, Abrüstung und internationale Sicherheit (Germany)
FY	Fiscal year
GMR	Giant magnetoresistance
GNR	Genetics, nanotechnology and robotics
GPS	Global Positioning System
HEMT	High electron mobility transistor
IBMT	Fraunhofer-Institut für Biomedizinische Technik (Germany)
ICRC	International Committee of the Red Cross
IR	Infrared
IRC	Interdisciplinary Research Centre
ISN	Institute for Soldier Nanotechnologies (MIT, USA)
ITRS	International Technology Roadmap for Semiconductors
LANL	Los Alamos National Laboratory (USA)
LLNL	Lawrence Livermore National Laboratory (USA)
MANPADS	Man-portable air defence system
MEMS	Micro-electromechanical system
MIT	Massachusetts Institute of Technology
MNT	Molecular nanotechnology
MoD	Ministry of Defence (UK)
MPU	Microprocessor unit
MST	Microsystems technology

NASA	National Air and Space Agency (USA)
NATO	North-Atlantic Treaty Organization
NBIC	Nanotechnology, Biotechnology, Information Technology and Cognitive Science
NEMS	Nano-electro-mechanical system
NIF	National Ignition Facility
NIH	National Institutes of Health (USA)
NIST	National Institute of Standards and Technology (USA)
NO	Nitrogen monoxide
NNI	National Nanotechnology Initiative (USA)
NRL	Naval Research Laboratory (USA)
NSET	Interagency Subcommittee on Nanoscale Science, Engineering and Technology (USA)
NSF	National Science Foundation (USA)
NSOM	Near-field scanning optical microscopy
NT	Nanotechnology/ies
ONR	Office of Naval Research (USA)
OSCE	Organization for Security and Co-operation in Europe
OTA	Office of Technology Assessment (USA)
PAL	Permissive action link
PBG	Photonic band gap
PC	Personal computer
PRIF	Peace Research Institute Frankfurt (Germany)
RAM	Random-access memory
RASCAL	Responsive Access, Small Cargo, Affordable Launch
R&D	Research and development
RDT&E	Research, development, testing and evaluation
RTO	Research and Technology Organization of NATO
SI	Système International d'Unités
SNL	Sandia National Laboratories (USA)
SPAWN	Satellite Protection and Warning
STM	Scanning tunnelling microscope
TAB	Büro für Technikfolgen-Abschätzung beim Deutschen Bundestag (Germany)
TACOM-ARDEC	Tank-automotive and Armaments Command – Army Research, Development and Engineering Center (USA)
UAV	Uninhabited air vehicle
UC	University of California (USA)
UHF	Ultra-high frequency
UK	United Kingdom
UN(O)	United Nations (Organization)

USA	United States of America
VCSEL	Vertical-cavity surface-emitting laser
WHO	World Health Organization
WTO	Warsaw Treaty Organization

1

INTRODUCTION

After Section 1.1 sets the stage, Section 1.2 presents the goals and an overview of this work. A short history of nanotechnology (NT) is given in Section 1.3. Section 1.4 lists promises and risks of NT. Section 1.5 relates previous writing on military uses of NT.

1.1 Nanotechnology: 'The next industrial revolution'[1]

In the coming decades, nanotechnology is expected to bring about a technological revolution. Working on the nanometre scale ($1\,nm = 10^{-9}m$ is a billionth of a metre), NT (including nanoscience) is about investigating as well as manipulating matter on the atomic and molecular level. At this level, the borders between the disciplines physics, chemistry, biology vanish, including their sub-, intermediate and applied fields, such as materials science, mechanics, electronics, biochemistry, genetics, neurology. NT is interdisciplinary and comprises many different areas; by making use of qualitatively different characteristics at the nanoscale, drastic miniaturization of components and utilization of molecular processes similar to those of life, NT is foreseen to lead to stronger but lighter materials, improved solar cells, markedly smaller computers with immensely increased speed and exhibiting general intelligence, micro and nano tools, large and small autonomous robots, great progress in molecular biology with the potential for medical intervention within cells, direct connections between electronic devices and nerve cells or the brain, etc. Expecting vast markets for NT products in the future, many countries are strongly increasing public funding for NT research and development; corporations are intensifying their efforts as well.

First products (e.g. nanolayered magnetic disk heads, nanostructured catalysts, nanoparticles in cosmetics) are already on the market. The great breakthroughs, however, are yet to come. Even if futuristic concepts of universal molecular assemblers and self-replicating nano-robots (so-called molecular NT) may not materialize, NT as it is foreseen today contains many far-reaching visions that would have vast impacts on individuals, societies and the international community.

1.2 Goals and overview of the study

Military uses of new technologies can create special problems and dangers – this holds all the more for NT. The goal of this study is to do a first assessment of the implications that NT weapons and other military NT systems could entail, and to present first considerations on preventive limitations.[2] It builds on my similar study of microsystems technology (MST) (Altmann 2001). Because NT comprises many diverse areas which are developing fast, they cannot be covered in detail. One important goal of this work is therefore to stake the ground for more detailed analyses of specific areas.

When I began investigations of NT (2000/2001, in-depth work started with the present project in early 2002), I had to rely on my own considerations for many potential military applications. In the meantime, military research and development programmes for nearly all of these have been announced in the USA, and I have referenced those. The US openness in military matters is unequalled, but has its limits, e.g. when it comes to weapons of mass destruction.

This study has the following structure: Chapter 1 gives an introduction to NT with its promises and risks and relates previous writing on military uses of NT. Chapter 2 gives an overview of NT with its various areas, research and development, and market outlook; a special Section (2.2) deals with assembler-based or 'molecular' NT. Chapter 3 describes the military efforts for NT in the USA and casts a short look at a few other countries. Potential military applications of NT are presented in Chapter 4. After a description of the concept and design of preventive arms control in Chapter 5, the military applications are discussed under criteria of preventive arms control in Chapter 6. The final Chapter 7 presents recommendations for political action as well as for further research and ends with concluding thoughts.

The reader will find some general NT literature in Appendix 1. Appendix 2 presents the NT-related research and development (R&D) efforts of the US Defense Advanced Research Projects Agency (DARPA), which is the largest military R&D programme world-wide, and the one where the most information is available.

1.3 Some NT history

Unknowing use of NT dates back thousands of years: nanoparticles of soot were used to produce ink already in ancient China, gold nanoparticles gave rise to the red colour in medieval stained-glass windows. Modern science 'arrived' at the nanoscale in one sense when the concepts of atoms and molecules were formalized and corroborated in the nineteenth century; in a more concrete sense, it did so when the first x-ray diffraction images of crystal structures were made and interpreted in the 1910s. Another important step was the (transmission) electron microscope (1930s) with which structures of nanometre size could be imaged. With the discovery of the atomic nucleus (1911) and later elementary particles (neutron 1932 and onward), physics research has moved on to femtometres (10^{-15}m $= 10^{-6}$nm) and below; however, at these scales the possibilities for stable structures, controlled manipulation and all the more for technical exploitation are quite limited due to electrostatic repulsion and quantum-mechanical effects.

Taken as a concept of manipulation below 100nm scale, NT is often traced back to the speech by R. Feynman of 1959, 'There's Plenty of Room at the Bottom: An Invitation to Enter a New Field of Physics', where he mentioned, among others, writing with nm-wide ion beams, computer components consisting of 100 atoms, production of small parts by a billion small factories (Feynman 1959). In the following decades, research and technology of the microscale made steady progress. A great step forward was the invention of the scanning tunnelling microscope (STM) in 1981 which allowed the first direct observation of single atoms in a surface, followed in 1986 by the atomic-force microscope (AFM). Only a few years later, these were used as tools to move single atoms around on a surface. The 1980s were also the period when the first articles and books by E. Drexler on molecular NT (MNT) appeared (Drexler 1981, 1986) and the Foresight Institute for MNT was founded.[3,4] In the 1990s and beyond, breakthroughs were achieved in many areas: carbon nanotubes (CNT) were discovered and a CNT transistor was demonstrated, DNA molecules were connected to form cubes and other three-dimensional structures, a single molecule acted as an electronic switch, an inorganic nanodevice was powered by a biomolecular rotating motor, molecular-dynamic computations were done of molecular planetary gears, kinesin molecules moving along microtubules were observed.[5] Recognizing the fundamental importance and wide range of future applications, the highly industrialized countries have greatly increased funding for NT R&D since the late 1990s. The founding in 2000 of the US National Nanotechnology Initiative (NNI) has much strengthened this trend and led to similar initiatives in many other countries (see Section 2.5).

1.4 Promises and risks of NT

1.4.1 Benefits of NT

Given the potential for structuring matter on the nanometre scale, NT is projected to deliver many benefits. The US NNI workshop on societal implications of NT of 2000/2001 mentioned (Roco and Bainbridge 2001: 3–11; see also Anton *et al.* 2001):

- lighter, stronger, more durable, programmable materials, allowing lighter vehicles,
- smaller, more powerful computers, sensors and displays,
- integration of biological with synthetic systems for pharmaceutical production,
- dramatically faster genome sequencing, individual therapeutics, targeted drug delivery,
- artificial materials for cell diagnostics, biocompatible implants,
- control and minimization of emissions from production, removal of contaminants from the environment,
- increased efficiency of solar-energy conversion,
- highly efficient fuel cells and hydrogen storage,
- nanostructured catalysts for chemical production with less energy and waste,
- nanostructured light-emitting diodes for saving energy in lighting,
- water purification and desalination,
- better chemicals for agriculture, genetic improvement for plants and animals,
- light-weight space launchers and spacecraft, miniaturized automatic space systems.

(For the promises of MNT, see Section 2.2.)

1.4.2 Risks of NT

Providing immense possibilities for applications in many areas, NT at the same time brings about a large potential for dangers, by negligence, accident or intentional action, directly and indirectly.

Direct risks can ensue from NT products or substances used in production. An urgent present problem exists with nanoparticles and nanofibres, production of which is being scaled up; however studies on their health and environment effects are widely lacking and first evidence exists that ultrafine particles, which owing to their smallness can penetrate body membranes, are not innocuous (ETC 2003a and refs; Howard C.V. 2003 and refs; see also Colvin 2003; Brumfiel 2003).

At the same 2000/2001 NNI workshop on societal implications, several risks and ethical problems of NT were mentioned, among them (Roco and Bainbridge 2001: 13–16; Weil 2001; Smith R.H. 2001; Suchman 2001; Meyer 2001; Tenner 2001; see also Anton *et al.* 2001):

- health risks for workers at new production processes,
- environmental problems from large-scale production, difficult-to-recycle nanocomposites,
- new products disturbing industries,

- glutted markets, upheaval of the global financial/manufacturing system,
- much reduced employment opportunities for less skilled labour,
- unequal distribution of benefits and wealth (e.g. in medicine), 'nano divide',
- equity disputes about intellectual property rights,
- conflicts of interest in university–industry relations,
- risks with genetic manipulation of plants, animals, humans,
- risks with implants,
- problems if diseases can be diagnosed long before cures become available,
- invisible intelligence gathering devices, covert activities,
- invasion of privacy, of human body without knowledge,
- security and safety of persons,
- superintelligent, virtually invisible devices from NT combined with artificial intelligence (AI),
- nanoweapons, artificial viruses, controlled biological/nerve agents,
- necessity for strict controls despite widening global knowledge.

The NNI workshop pointed out that unintended indirect – good or bad – consequences have to be expected; e.g. a longer life expectancy would require changes in pensions or retirement age; nanoparticles in the environment could lead to extensive biological change as with DDT, but finding that out and adjusting may take a very long time.

To deal with such risks, the workshop recommended that the scientific, technological and societal impacts and implications be studied in systematic, interdisciplinary research. In particular, social-science research should be done with high priority (Roco and Bainbridge 2001: 20–24). The 2002 European/US NT workshop has argued similarly (Roco and Tomellini 2002: 21–24). However, a review of the US NNI in 2002 found only little societal-implications research and recommended a new funding strategy (NNI Committee 2002: 34–35, 48–49). A group of medical and bioethics researchers from Canada observed a 'paucity of serious, published research into the ethical, legal, and social implications of NT' (Mnyusiwalla *et al.* 2003). An article by the NNI co-ordinator has described the NNI activities, but has not discussed societal implications that might follow from military uses – even though an international perspective is explicitly mentioned and despite the fact that more than a quarter of the NNI funds goes to the US Department of Defense (see Section 3.1.1) (Roco 2003a).

Motivated by the risks, non-governmental organizations have begun to care about NT; the first longer studies were published by the ETC Group (ETC 2003, 2003a)[6] and Greenpeace (Arnall 2003).

On the official side, parliamentary and governmental institutions have launched studies in NT. In Germany, the Federal Ministry of Education and Research (BMBF) tasked a preliminary study in 2000 (Malanowski 2001; see also Bachmann and Zweck 2001). The Office for Technology Assessment at the Federal Parliament (TAB) carried out a major project in 2002–2003 (Paschen *et al.* 2003).[7] The US House of Representatives held a hearing in April 2003 (House 2003). In the UK, the Better Regulation Task Force advising the Government in January 2003 singled out NT, recommending openness, informed public debate and a strong government lead on risk issues (BRTF 2003). In June 2003, the UK Department of Trade and Industry (Office for Science and Technology) commissioned a study on benefits and possible problems of NT to be carried out by the Royal Society and the Royal Academy of Engineering (Royal 2003). In December 2003, the US Congress decided to establish an 'American Nanotechnology Preparedness Center' to work on 'societal, ethical, environmental, educational, legal, and workforce implications' of NT and 'anticipated issues related to the responsible research, development, and application' of NT (Congress 2003). Obviously, NT and its implications are taken seriously by governments and parliaments.

1.4.3 Risks of MNT

Concerning MNT, its major general risks were already referred to by Drexler in 1986. Proceeding from his concept of self-replicating nanomachines, he mentioned the so-called 'grey-goo' problem, omnivorous replicators consuming all organic material on Earth, and noted some ideas on limiting the capabilities of assemblers.[8] Later, the Foresight Institute took up this thread and published principles and design guidelines for self-replicating devices (Foresight 2000).

On the societal level, several other risks of MNT have been mentioned (CRN 2003). Economic disruption could result from cheap products and large-scale displacement of human labour. MNT products might nevertheless be too expensive for the very poor, increasing the inequalities in the world, potentially leading to social unrest. Ubiquitous small sensors with powerful computers could be used for continuous surveillance of all citizens. Cheap devices could be used for physical or psychiatric control.

The warnings of computer scientist B. Joy have become widely known and discussed (Joy 2000).[9] Citing genetics, NT and robotics as the powerful technologies of the twenty-first century, he mentions their dangers, among them robots succeeding humans and knowledge-enabled mass destruction, hugely amplified by self-replication. Too powerful to be contained by 'shields', these technologies need to be relinquished; their development and pursuit of certain kinds of knowledge are to be limited.

Reactions from the MNT community have stressed the problems following from relinquishment and over-regulation (e.g. Reynolds 2002).

Recently, a Center for Responsible Nanotechnology has been founded by two people from the MNT community; they give a wide list of MNT risks and argue for moderate restrictions (CRN 2003).

One should note that in the MNT community, concepts of transcending human existence, immortality, colonization of space etc. are generally not seen as problematic, but rather advanced as beneficial (see Section 2.2).

1.5 Previous writing on military uses of NT

Given its far-reaching potential, NT can have extensive effects on warfare and the armed forces. Up to now, however, there has been little literature on military uses of NT, but many of the basic implications have been long understood, and were satirized by Stanislaw Lem as early as 1983 (Lem 1983). Most contributions have been inspired by the concept of MNT. After the US NNI was founded and incorporated national security and defence issues on a high level, its focus was rather on medium-term implications of NT, but with openness towards revolutionary changes.

In this section, a selection of writings is related that have discussed military uses of NT on a general level or have mentioned arms-control aspects of NT. Where appropriate, short comments are given. Detailed, technical texts from brainstorming or planning for the military, such as the US Army Workshop on Nanoscience for the Soldier (ARO Nanoscience 2001),[10] are referred to in Chapters 3 and 4.

1.5.1 E. Drexler 1981–1991

When K. Eric Drexler presented his concept of 'molecular engineering' in 1981, he mentioned opportunities and dangers in one sentence at the end (Drexler 1981). In his 1986 book, *Engines of Creation*, where he presented the NT concept to the general public, he had one chapter on 'Engines of destruction' and one on 'Strategies and survival' (Drexler 1986/1990: Chs 11, 12). Beside the general dangers of the so-called 'grey-goo' problem and the displacement of humans, he mentioned the intentional use of replicators and AI systems for military power: the former for building large numbers of advanced weapons or waging a sort of germ warfare, and the latter for weapons design, strategy or fighting. Based on the assembler breakthrough, a state could rapidly extend its military capabilities – which would lead to sudden and destabilizing changes.[11] Because replicators would not use rare isotopes and could start with a very small amount of material, they could be more potent than nuclear weapons in leading to extinction. On the other hand, nanomachines could be used much more flexibly than bombs, e.g. within oppressive states, for spying or for body manipulation. In order to delay creation of dangerous replicators, Drexler recommended that the leading force take measures such as limiting and sealing assemblers, and hiding information. In the long run, however, an 'active shield' would be required, that is automated defensive nanomachines which fight dangerous replicators of all sorts. Writing during the Cold War, he recommended international treaties and co-operation with the Soviet Union, but with the technological lead of the democracies. Traditional arms control based on verifiable limitations would not likely be able to cope with NT. Active shields would provide protection without threatening.

In their subsequent book of 1991, E. Drexler and C. Peterson again devoted two chapters to risks of molecular NT and policy measures (Drexler and Peterson 1991: Chs 12, 13). They argued that with appropriate precautions, accidents and unintended consequences could be limited. The chief danger would stem from abuse, or intentional use for destructive purposes. Because of the terrifying prospects of an NT arms race, international arms control through co-operative development should look attractive, but this would not be easy or likely. Regulation would buy time, but ultimately protective technology would be needed against novel nanoweaponry. This is explicitly called the greatest challenge of any the authors have discussed, and in an aside they state that their outlook is not an optimistic one. Suppression of research is not seen as sensible, because not all countries would comply, and NT research would only be pushed to secret military work. Of the five scenarios described,[12] only one is suitable to avoid catastrophe; it is characterized by international partnership of the democracies, regulation of technology transfer, domination of economic co-operation over military competition, due attention to potential military threats, and 'mutual inspection' which comes with co-operative R&D.

It is obvious that the concept of 'active shields' or 'protective technology' is a fairly diffuse one. It is by no means clear why defensive MNT systems should be superior to offensive ones, and the former could bring (or evolve into) severe dangers of their own.

1.5.2 D. Jeremiah 1995

At the 4th Conference on Molecular Nanotechnology of the Foresight Institute in 1995, Admiral David E. Jeremiah (US Navy, retired, former Vice Chairman of the Joint Chiefs of Staff) gave a speech on 'Nanotechnology and Global Security' (Jeremiah 1995). He stressed several general reasons for armed conflict in the future, such as ethnic strife, competition over resources, population and environment problems, migration, the technological revolution. Emphasizing the growing military importance of information, he foresaw miniature sensors scattered in very high numbers, implant enhancements in the human body, and domination of combatants and weapons by robotics. He speculated that MNT, together with AI, could lead to humanoid robots with artificial brains for nations with reduced manpower or an aversion to bloodletting. He posed the problem how mischief with molecular NT could be avoided, given that the effects could be greater than with nuclear weapons, and warned against restrictions that would give an advantage for those operating outside of them.

1.5.3 M. Gubrud 1997

A less enthusiastic view of military MNT was presented by Mark A. Gubrud in 1997 (Gubrud 1997).[13] He discussed what would happen to international security if self-replicating universal assemblers came into being in our world consisting of nation states preparing for armed conflict. MNT would not only allow greatly improved conventional weapons or miniaturized combat systems in the sea, but massive military production on a very short time scale, including nuclear weapons. Using the revolutionary changes in quality and quantity, a nation with sufficient lead could in theory disarm potential competitors. To prevent this, nations would engage in an arms race characterized by sudden breakthroughs and new threats. Automated production using mainly local resources would reduce international trade and its source of common interest; the world system based on wage labour, transnational capitalism and global markets would vanish; a golden age would be possible, but economic insecurity, inequality, etc. could lead to political chaos. Competition over newly accessible resources in the oceans or in outer space could generate hostility even between democracies. Against limited numbers of nuclear weapons, MNT could provide for efficient civil defence, active defence and counterforce weapons undermining deterrence, allowing for victory over a major power. As a counter, MNT could be used for a massive nuclear-weapons build-up to hundreds of thousands or millions of warheads.

Beside arms-race and first-strike instability, MNT would bring a new type, early-advantage instability: in a scenario of exponential growth of military production (before resource limitations became effective) with characteristic times of hours or days, starting earlier could mean a decisive difference in force level. Gubrud stated that a US lead would not last, since the industrial technology base would be similar in other countries. Progress to advanced generalized artificial intelligence (AI) of human-like capability or beyond would lead to similar dilemmas, even without MNT. Combined, both would greatly exacerbate the dangers.

In order to avoid a catastrophic MNT arms race, Gubrud recommended several arms-control measures. Not producing arms in masses and not preparing facilities for this would be verifiable, augmented by voluntary transparency. Most important at present would be a ban on space weapons, and continuing and completing nuclear disarmament. Nations should surrender some sovereignty to permit intrusive verification and engage in military co-operation. Ultimately, a single integrated global security system would be needed.

1.5.4 L. Henley 1999

In a US Army journal, Lonnie D. Henley wrote about the next military-technological revolution, based on the convergence of information processing, biological sciences and advanced manufacturing techniques such as NT (Henley 1999). Biological molecular processes could provide role models for 'wet' NT, different from the 'dry' version envisioning wheels, gears and electronics. Concerning military applications, Henley cautioned that the first useful products could be available in twenty years. He listed several potential capabilities, among them: selective biological weapons that can be triggered or that kill or incapacitate in a new way; decentralized nets of small sensors, down to 'surveillance dust'; improved information processing modelled after the human brain; fleets of small, inexpensive robots for attack. With a new miniature or microscopic theatre of combat, defence may be much more difficult than offence; continuing development of new biological-warfare agents would require a constant struggle in peacetime to neutralize enemy efforts. If biological processes are used to produce devices, it will be tempting to use reproduction – which would lead to escape and uncontrolled evolution with unexpected and perhaps adverse effects. Henley had no doubts that the USA will remain at the cutting edge commercially and scientifically. He asked if the USA will take the lead in this other vision of futuristic warfare.

1.5.5 B. Joy 2000

In his often-cited article 'Why the future doesn't need us', B. Joy warned mainly of the general dangers from genetics, NT and robotics (GNR) (Joy 2000). Concerning military and terrorist uses, quoting Drexler, Joy mentions massively, but selectively destructive NT devices. Whereas the nuclear, biological and chemical technologies used in the weapons of mass destruction of the twentieth century were largely military and developed in government laboratories, Joy writes, the GNR technologies of the twenty-first century are mostly commercial and are developed within corporations. Joy cites the nuclear arms race and quotes authors who see the chances for extinction of humankind from all the dangerous techniques at 30–50 per cent. Drexler's active NT shield against dangerous replicators 'would itself be extremely dangerous – nothing could prevent it from developing autoimmune problems and attacking the biosphere itself'. Similar problems would exist with shields against robotics and genetic engineering. A basis for hope is in the relinquishment of the USA of biological and chemical weapons within the respective international conventions. Building on this, abolition of nuclear weapons could help to achieve relinquishment of certain dangerous GNR technologies, including the commercial sector. Verification would be difficult, but not unsolvable – similar to that needed for biological weapons, but on an unprecedented scale, raising tensions with individual privacy and the desire for proprietary information. Scientists and engineers would need to adopt a strong code of ethical conduct, and to have the courage to blow the whistle if necessary.

1.5.6 Foresight Guidelines 2000

The Foresight Institute (Palo Alto CA, USA),[14] founded in 1986 by Eric Drexler and others to promote MNT, has discussed dangers and misuse of self-replicating artificial systems, and has developed guidelines to prevent or minimize these risks (Version 3.7 of 4 June 2000, Foresight 2000). The guidelines deal with regulation on the national level, and propose technical as well as administrative measures to prevent uncontrolled replication in a natural environment. The guidelines ask for the development of efficient means of restricting misuse in the international arena. They even discuss the option of including molecular NT into existing chemical-, biological- or nuclear-weapons treaties, but dismiss this option flatly. The argument is that 'a 99.99% effective ban would result in development and deployment by the 0.01% that evaded and ignored the ban' which would lead to economic and military disadvantages for the USA and other compliant states.

Even though molecular NT is not an imminent reality, it is important to state that this approach is clearly flawed in several respects. First, foregoing international regulation will have the consequence that all technologically capable countries will work actively for military applications of molecular NT as soon as that will appear within reach. Second, the argument neglects that technological capabilities are not evenly distributed – most of the countries that the USA sees as the strongest threats are not leading in high technology. Third, exactly 100 per cent verification is never achievable in practice; following this logic, one could not conclude a single arms-limitation treaty. For this problem, the notion of 'adequate' verification has been in use for several decades – the relevant consideration is whether the risk from potential undetected cheating is higher than from not having limits at all. The required degree of verification can be assessed on this basis for each area of military systems or technology. Admittedly, compliance with limits on molecular NT would be difficult to verify, but new technologies can be applied to verification as they become available. For early developments, strong similarities exist with biological agents. Thus, the verification protocol developed for the Biological Weapons Convention with its complex mix of inspections rights and protection of confidential information ('managed access' etc.) provides a good starting point (Feakes and Littlewood 2002; BWC AHG 2001).

1.5.7 S. Metz 2000

In a general article on the future of technology and war, Steven Metz mentioned the trend of miniaturization, with NT as a continuation of MST (Metz 2000). MST and NT could lead to things like a 'robotic tick' which could attach itself to an enemy system, then gather information or perform sabotage at a certain time. He cites Libicki's three stages of future warfare enabled by cheap information processing (Libicki 1994), where the third, 'fire-ant warfare', uses swarms of many small, relatively simple weapons. Quoting other references (Commission 1999; Robotics 1997; Henley 1999 (see Section 1.5.4); Drexler and Peterson 1991), Metz writes that – as the beginnings of cyborg capabilities – soon sensors and other systems might be mounted on dogs, rats, insects or birds, steered by an implant; MNT with molecular biology and information science could lead to selective biological warfare agents that can be triggered. Following this, he raises the strategic, operational and ethical issues connected with robots that kill.[15]

1.5.8 NNI Workshop 2000/2001

The second chapter of the first workshop on societal implications of NT organized in the context of the US NNI (September 2000) is devoted to NT goals, from understanding of nature via medicine, sustainability and space exploration to moving into the market. Under 'National Defense', the following applications are given (Roco and Bainbridge 2001: Ch. 2; similar in NNI 2002: Ch. 4):

1 continued information dominance through advanced electronics,
2 more sophisticated virtual reality systems based on nanostructured electronics enabling more affordable, effective training,
3 increased use of enhanced automation and robotics to offset reductions in military manpower, reduce risk to troops and improve vehicle performance,
4 higher performance (lighter weight, higher strength) in military platforms with diminished failure rates and lower life-cycle costs,
5 improvements in chemical/biological/nuclear sensing and in casualty care,
6 design improvement in systems for nuclear non-proliferation monitoring and management, and
7 combined nano- and micromechanical devices for control of nuclear defence systems.

Chapter 3 on 'NT and societal interactions' discusses unintended and second-order consequences, e.g. a higher inequality in the distribution of wealth (the 'nano divide') and environmental pollution (Roco and Bainbridge 2001: Ch. 3; similar in recommendations, Ch. 5). For assessment, the editors demand an examination of the entire system through its entire life cycle, with the participation of social scientists to allow for early identification of important issues and corrective action. They ask for ethical considerations to be included not only on the part of philosophers etc., but also by incorporation into the curriculum of scientists, technologists and technicians, to strengthen their individual responsibility in generating powerful new nanotechnologies. Military applications of NT and their effects, e.g. on peace and stability in the international system, are, unfortunately, not mentioned here.

The contribution by Yonas and Picraux (2001) gives a different point of view. They argue that national security will have to be sought in a context of global security. For the military, NT will enhance situational awareness by faster computers, better sensors and communication, and swarms of small, smart robots. They state that such advances could contribute to global stability in peacekeeping or resisting aggression, but point out that they could also further military aggression, and that the technologies – if cheap and widely available – could expand threats from terrorists or paramilitary groups. Solutions to this problem are not discussed.

Writing about implications for knowledge and understanding, Whitesides and Love (2001) present a short paragraph on national security. They mention quantum computing for cryptography and – in the context of war fighting with US technological superiority, with minimized casualties – the need to move and analyse staggering amounts of information for global information systems.

The contribution by Tolles (2001) is dedicated to national-security aspects of NT. After asking for a balance in economic and military efforts, he stresses that NT will lead to evolutionary improvements in military technology, warns of advantages open to potential adversaries, and lists opportunities for technological superiority: higher-performance platforms, enhanced sensing, enhanced human performance, improved processing and communication, safer operation in hazardous circumstances through remotely operated robots, reduced manpower through greater automation, improved casualty care, remediation of chemically or biologically contaminated areas or equipment and lower life-cycle costs through improved materials and condition-based maintenance. Turning to visionary claims about molecular assemblers, self-replicating machines, computers out of human control etc., he calls for a rational approach, avoiding misinformation. On the one hand, Tolles calls for a strong NT defence programme; on the other hand, he notes that NT can provide enhanced resources for the world which tend to reduce tensions and to increase national security for all. Potential conflicts between these two goals are not taken into account.

1.5.9 UK Ministry of Defence 2001

In 2001, the UK Ministry of Defence (MoD) published a discussion paper, 'The Future Strategic Context', which looks forward up to 2030 in seven dimensions, among them technological, economic, political and military (UK MoD 2001). In the technological dimension, the paper mentions as possible developments by the end of the period: computers with increasing processing power allowing new applications, propulsion and power generation enabling long-endurance remotely deployed systems and micro unmanned airborne vehicles, direct and indirect electronic-brain links with implanted and surface/remote equipment, machines capable of autonomous intelligent judgements, genetics allowing new forms of biogenetic warfare or terrorism. In NT proper, mention is made of miniaturization of sensors and equipment, nano-solar cells, micro-platforms for reconnaissance and nano-robots for many purposes, including medical robots fighting diseases or repairing DNA internally in humans.[16] Military doctrine etc. would have to change in parallel to revolutionary changes in technology. Military R&D should emphasize the relevant areas not interesting for the civil sector, closely monitor advances in other countries, preserve the edge in key areas, exploit civil R&D and ensure access to technology of other nations, in particular the USA.

More concrete ideas were presented in an information sheet on NT and its impact on defence and the MoD (UK MoD 2001a). NT could have enormous implications – opportunities for the country's own military capability as well as new threats. A non-exhaustive list of possibilities reads:

- completely secure messaging,
- intelligent and completely autonomous short- and long-range highly accurate weapons,
- improved stealth but also means to defeat current stealth techniques,
- global information networks and local battlefield systems with 'all-seeing' sensors,
- miniature high energy battery and power supplies,
- intelligent decision aids,
- self repairing military equipment,
- new vaccines and medical treatments,
- highly sensitive miniature multiple biological and chemical sensors,
- unethical use leading to new biological and chemical weapons.

According to the MoD, 'it is unlikely that the human will be taken out of the loop for key decisions like weapons release'. Since much of the basic technology will arrive in the civil sector, it will also be available to would-be enemies who might utilize it faster because of simpler acquisition systems. Given the emphasis on high technology in the UK military, this statement does not seem convincing. However, the warning of new threats from terrorist attacks using NT-manipulated biological and chemical species is certainly justified.

1.5.10 NBIC Convergence Workshop 2001/2002

The thematic group on National Security of the US-government-sponsored workshop 'Converging Technologies for Improving Human Performance Performance – Nanotechnology, Biotechnology, Information Technology, and Cognitive Science' (NBIC) of December 2001 (see Section 2.3) did not go as far in its visions. It identified seven goals for integrated NBIC (Roco and Bainbridge 2003: Section E):[17]

1. Data linkage, threat anticipation and readiness (miniature sensors, high-speed processing and communication).
2. Uninhabited combat vehicles (air vehicles with artificial brains emulating a skilful pilot, similar for tanks, submarines etc.).
3. Warfighter education and training (inexpensive, high-performance virtual-reality computerized teaching, with speech, vision and motion interaction).
4. Chemical/biological/radiological/explosive detection and protection (micro sensor suites, protective masks and clothing, environmentally benign decontamination, physiological monitors and prophylaxis).
5. Warfighter systems (electronics with 100 times memory size and processing rates, flexible, thin displays or direct write onto retina, netted communication, weapons tracking targets, physiological monitors for alertness, chemical/biological agents, and casualty assessment; small volume, weight and power).
6. Non-drug treatments for enhancement of human performance (modify human biochemistry – compensate for sleep deprivation, enhance physical and psychological performance and survival rates from injury).
7. Applications of brain-machine interface (take brain signals nonintrusively, use with feedback for control of systems).

All this, it is claimed, will provide the USA with an overwhelming technological advantage, reducing the likelihood of war. Technological arms races, potential destabilizing developments or new criminal threats are not taken into account.

1.5.11 Petersen and Egan 2002

In March 2002, the Center for Technology and National Security Policy of the US National Defense University, Washington DC, published an overview article on NT and future defence (Petersen and Egan 2002). After a short overview of NT and some of its promises, defence applications are discussed. The article does not differentiate between mid-term goals as expressed in the NNI (see Section 1.5.8) and visionary NT, such as nano-robots, a coming 'singularity', etc. Echoing warnings of instability by Gubrud (see Section 1.5.3), the authors argue for improving global security by reducing the economic disparity, using the military rather to prevent conflict than to fight wars.

1.5.12 Air Force Science and Technology Board 2002

The Air Force Science and Technology Board of the US National Research Council formed a committee to study the implications of micro- and nanotechnologies for the Air Force; its 240-page report was presented in 2002 (NRC Committee 2002). Named as overarching themes are: increased information capabilities, miniaturization, new engineered materials, increased functionality and autonomy. In view of extensive efforts world-wide, Air Force R&D should be selective and coupled to the former. A detailed taxonomy of MST/NT opportunities has been developed; they are relevant for all Air Force core competencies. Findings and recommendations are offered for technology as well as policy. In technology, keywords are: miniaturization of electronics, new materials, biological science (biomimetics, biocomputing, enhanced human performance), fixed arrays and moving swarms of sensors, propulsion and aerodynamic control, co-design of hardware and software, self- and directed assembly. With respect to policy, R&D funding should be increased, work should be concentrated on basic research and on Air-Force-specific applications, with good planning and co-ordination.

1.5.13 S. Howard 2002

In summer 2002, the journal *Disarmament Diplomacy* carried two articles discussing NT and weapons of mass destruction. In the first, Sean Howard, after a short look at the history of NT, the military share of the US NNI and the new Institute for Soldier Nanotechnologies (see Section 3.1.6), relates Drexler's grey-goo scenario and defends Bill Joy's warnings of NT-enhanced chemical and biological weapons (see Section 1.5.5) (Howard S. 2002). Similarly to nuclear physics in the 1930s when physicists were sceptical of nuclear energy release, continued NT research would at some time make feasible new weapons of mass destruction which could then be used by terrorists. In order to prevent this, he discusses the options of abolition versus regulation of what he calls 'inner-space research'. For both options, Howard presents sketches for draft texts of an 'Inner Space Treaty'. The first declares inner (atomic and molecular) space free for exploration and engineering by all states. Atomically engineered objects with weapons of mass destruction are banned, as is any type of nanotechnological weapon. The second Treaty version would completely prohibit 'any activities ... relating to the nanotechnological exploration and engineering of inner (atomic and molecular) space'. It declares inner space an International Protectorate.

The underlying goal – preventing NT-enabled new types of weapons – is certainly laudable. The proposed measure of dealing with the problem, however, seems grossly inappropriate. Both draft Treaty texts suffer from the same basic problem: they take the wording of the Outer Space Treaty of 1967 and convert it to the 'inner space'. Thus sentences result of which it is difficult to make sense. For example, what does it mean that 'there shall be free access to all areas of the nanosphere', that 'inner space is not subject to national appropriation', that 'States ... shall carry on nanotechnological activities in the exploration and engineering of inner space in accordance with international law'? Is it sensible to oblige states to inform the UN Secretary General of NT exploration activities and oblige the latter to disseminate this information? All such concepts neglect the many fundamental differences between outer and inner space. The former is vast, and its exploration needs great funds, so that only few activities are taking place. Outer-space activities, in particular launches, are observable from thousands of kilometres, making non-intrusive verification a relatively simple matter. 'Inner space' is available everywhere, so that investigations and engineering can occur at many places; verification would be very costly and very intrusive. The stipulation that 'all nanotechnical facilities ... shall be open to inspection' could make sense, but one would want a thorough discussion of its consequences – these would of course be much more severe than the corresponding rule for stations on the moon. The main flaw of the abolition version is that it does not address the research of atomic and molecular space that has been done over the last century, and the technology based upon it, e.g. elementary-particle accelerators, discovery of the DNA structure, microelectronics, molecular biology. It seems that concrete, differentiated and detailed considerations of science and technology cannot be avoided even if one wants to apply fairly wide-ranging limitations.

1.5.14 A. Gsponer 2002

The second article in *Disarmament Diplomacy* treats NT as it relates to so-called fourth-generation nuclear weapons (Gsponer 2002). The article is fairly general, and some of its allegations are questionable.[18] The main point is that MST and NT can provide the possibility for pure-fusion, thermonuclear explosions of very small yield (tons, not kilo- or megatons of TNT equivalent), in devices of only a few kilograms mass.

It is true that, if such weapons come into being, the strategic situation would change drastically, with a blurred distinction between conventional weapons and weapons of mass destruction. However, it is not clear how the success of fusion ignition in micropellets using giant lasers in large halls, as in the US National Ignition Facility (NIF),[19] could be transferred to kilogram-size devices. Tracking scientific-technical progress in the area of advanced fusion triggers is certainly needed; for more discussion see Section 4.1.19.2.

2

OVERVIEW OF NANOTECHNOLOGY[1]

Nanotechnology (NT) is defined and introduced in Section 2.1. The still futuristic molecular NT with its associated concepts is discussed in Section 2.2. Section 2.3 describes the trend towards convergence of nano-, bio-, information and cognitive sciences and technology. Section 2.4 presents the areas of NT, and Sections 2.5 and 2.6 look briefly at R&D of NT and the expected market, respectively.

2.1 General aspects

2.1.1 Definition

Nanotechnology derives its name from a nanometre, a billionth of a metre ($1\,nm = 10^{-9}\,m$). Nanotechnology deals with structures the sizes of which are between about 0.1 nm (single atom) and about 100 nm (large molecule).[2] Sometimes the term nanoscience is used for research at the nanometre size scale whereas nanotechnology has its focus on artifacts; usually, however, NT is understood in a wide sense to encompass research, not the least because much research is directed to applications. The same wide notion of NT is used here.[3]

Above NT size, from $0.1\,\mu m$ (=100 nm) to several $100\,\mu m$, the term microsystems technology (MST) is often used;[4] here, microelectronics came first and constitute a predominant area. Systems between micro- and macro (decimetres to metres) size are sometimes called mesoscopic.[5] Table 2.1 gives a few typical sizes.

In one sense, NT is the logical extension of MST to a smaller scale, and the development of microelectronics and MST with its drive to further miniaturization is naturally leading to sizes below 100 nm (see Section 2.1.3). On the other hand, NT opens up many new areas where qualitatively new phenomena occur. MST and NT are mutually reinforcing, and the divide between them is fuzzy. MST provides tools – mediators to the macro world – such as cantilever probes, for analysis and manipulation at the nanoscale (e.g. MacDonald 1999). NT provides new application areas

Table 2.1 Typical sizes for comparison. NT comprises the size range from about 0.1 to about 100 nanometres (nm). 1 micrometre (μm) = 10^{-6} m = 1000 nm; 1 nm = 10^{-9} m; 1 picometre (pm) = 10^{-12} m = 0.001 nm; 1 femtometre (fm) = 10^{-15} m = 0.000,001 nm

Atomic nucleus	1–7 fm
Silicon atom (in crystal)	0.24 nm
Water molecule (largest diameter)	0.37 nm
Carbon nanotube (diameter)	0.7–3 nm
DNA molecule, width	2 nm
Protein molecule (hemoglobin, diameter)	6 nm
Transistor in modern integrated circuit	100 nm
Animal cell (diameter)	2–20 μm
Human hair (diameter)	50–100 μm

for MST, e.g. for DNA recognition using DNA fragments with different patterns specifically bound in the microfabricated pits of a silicon chip. Combinations of MST and NT will lead to a greatly expanded spectrum of applications.

2.1.2 Top-down versus bottom-up production

The traditional way of producing small structures is *top-down*: e.g. in microelectronics and MST, macroscale equipment is used to remove or add material with very high precision. Scaling down to the nanoscale meets technical problems: smaller mask structures require correspondingly smaller wavelengths of the radiation used to expose the photoresist; soon, ultraviolet light will arrive at its limits. Increasing requirements on mechanical precision and cleanliness lead to greatly increasing costs – the cost of a semiconductor fabrication line crossed the $1-billion mark in the mid-1990s (Timp *et al.* 1999: Section I A). Some nanosystems could, in principle, be built by moving atoms and small molecules on a surface using a scanning tunnelling microscope, or by exposing surfaces to beams of ions or electrons which have very small wavelengths (on the order of 0.05 nm). This is of course very tedious and produces only one system at a time.

Cheap production requires making many similar systems in parallel. One way out is to do the slow structuring only once on a master, and then use simple processes to produce many copies, as in lithography, stamping, embossing or printing. Another approach to avoid costly manipulation of single nanosystems is the use of many automatic nano/micromanipulators that work in parallel.[6]

A fundamentally different alternative is to produce *bottom up* where components arrange themselves by physical/chemical forces. Examples of such self-organization or self-assembly are quantum dots forming two-dimensional arrays on a semiconductor substrate of different lattice parameter, or alcane-thiol molecules the sulphur ends of which adhere to a gold surface, forming a monolayer with parallel orientation of the molecules (e.g. De Wild *et al.* 2003). These modes of self-organization are relatively simple. Forming complex structures, such as three-dimensional computing/storage elements and their connections, poses much higher requirements. Biological processes as in nerve growth are inspirational models here but will be difficult to reproduce in artificial systems.

The final goal of NT is the control of the type and three-dimensional position of each single atom in a molecule or a larger structural unit – 'shaping the world atom by atom', as the brochure of the US NNI puts it (IWGN 1999a).[7] In such a way, materials and systems are to be produced that are difficult or impossible to make by traditional chemical, i.e., thermodynamic or stochastic, processes. The only limits are set by the fundamental laws of nature. Doing this economically will require bottom-up processes – from simple self-assembly to micro- and nanomachinery and life-like growth.

2.1.3 From micro- to nanoelectronics

A prominent special case is provided by the continuing miniaturization of electronics where extrapolation of the observed scaling trends is now leading to NT (e.g. Timp *et al.* 1999). For about four decades, Moore's Law has held: the number of components per integrated circuit doubles about every 1.5 years. This has been achieved by continuous down-sizing of components; in parallel, clock rates and chip sizes have increased. According to the averaged trend, in 2003 the characteristic feature size is 100 nm, and 90 nm is foreseen for 2004 – thus we are just now entering the size range of the NT definition. In the 2001 roadmap of the semiconductor industry, extrapolation continues through 2016 with a feature size of 22 nm, Figure 2.1. Achieving such a rate of down-sizing will require overcoming significant barriers in the coming years. After 2008, next-generation lithography (such as extreme ultraviolet, electron projection) will be needed to achieve the 45-nm node; new solutions for interconnects might include wireless or optical principles. In parallel to the shrinking size, factory costs are escalating. Further continuation of the trend will require using different principles. Research is going on in several different directions: magnetic or ferroelectric memory, single-electron transistors, crossing nanowires, organic molecules as conductors, for memory and as switching elements, 'writing' with nanoscale probes on surfaces, using the third dimension (Timp *et al.* 1999 and refs; Reed and Tour 2000; Gracias *et al.* 2000; Lieber 2001; Vettiger *et al.* 2002; Chen *et al.* 2003 (for doubts about the latter see Service 2003)). Top-down structuring would likely lead to very slow processes and prohibitive costs. Major challenges lie in finding bottom-up processes of forming the elements and their interconnections.

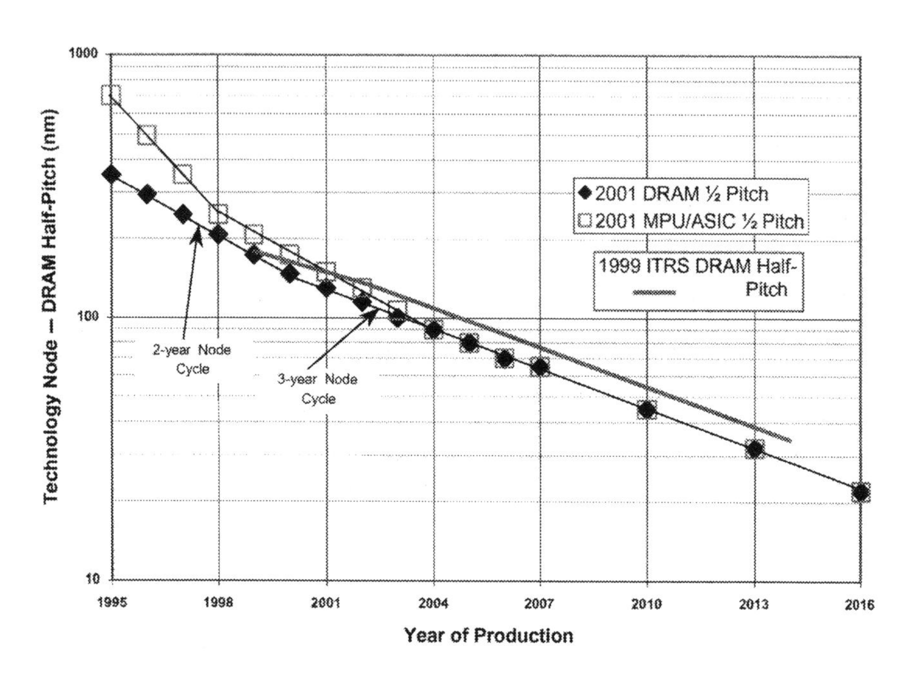

Figure 2.1 Shrinking of characteristic features of integrated circuits according to the International Technology Roadmap for Semiconductors (ITRS) of the semiconductor industry. From 2004, the half pitch[8] of processing circuits (microprocessor units (MPU) and application-specific integrated circuits (ASIC)) will be the same as that of dynamic random-access memory (DRAM) chips. Generations (technology nodes) do not change every year, the points are interpolated. Compared to the 1999 roadmap, predictions have been accelerated by one year, but the 2003 issue noted the possibility of a one-year delay (ITRS 2003: 38). Note that the size axis is logarithmic. (From SIA 2001: 34, free for public use.)

2.1.4 New effects at the nanoscale, new material properties

Structures below a few times 10nm size exhibit different properties than larger ones. This is due to several effects. One is geometric: the ratio between surface and volume (or mass) grows as the size shrinks. Interface areas for adsorption or chemical reaction increase if smaller particles or pores are used. This effect can greatly improve catalysis, storage density for fuels (e.g. hydrogen) and reaction efficiency and power density of energy conversion (e.g. in fuel-cell membranes). In bulk materials, smaller crystallites can improve mechanical properties. Considering scaling of forces influencing motion, friction may become less relevant while adhesion may become more relevant.

Second, at nm size new quantum-mechanical effects often dominate. In small particles, the electrons can no longer occupy continuous energy bands as in larger solid bodies, but are confined to a few narrow levels that depend on particle size. This can change the electrical conductivity radically: it becomes quantized – depending on the occupation of the energy levels electrons may have to enter and leave a conductor singly. In somewhat larger structures, the collective flow of electrons may take place in the so-called ballistic mode, much faster than thermal equilibrium velocity, with nearly no scattering, and accordingly low ohmic losses and the potential for very high current densities and switching rates.

Because the energy levels determine the optical properties, absorption and emission of light can be tuned by the material and size of the particles, e.g. the fluorescence of semiconductor quantum dots from the visible to the near infrared.[9] A special aspect is collective excitations of electrons in surface states, so-called surface plasmons; with gold nanoparticles, the plasmon resonance gives rise to the red colour used in medieval stained glass.

Using very specific molecular structures, particular effects can be produced. Many examples are provided by biological systems, e.g. conversion of chemical energy into motion (kinesin, myosin), absorption of light quanta with energy transferred to another site for storage or signalling (rhodopsin in bacteria or eye sensory cells).

2.1.5 Carbon nanotubes: tools for many purposes

Carbon nanotubes, discovered in 1991, have become a focus of NT R&D (Goronkin *et al.* 1999; Busbee 2002: 9–11; Moriarty 2001; Salvetat-Delmotte and Rubio 2002; Lau and Hui 2002; Meyyappan and Srivastava 2002). They are one to a few nm thick and have been produced with lengths of several μm, with single or multiple walls. The simple form with a single wall can be conceived of as a single layer of a graphite crystal rolled up to a cylinder, with fullerene half-spheres closing the ends (Figure 2.2). Depending on the diameter and chirality, single-wall carbon nanotubes are metallic or semiconducting. Multi-wall nanotubes consist of many layers and have less distinct electrical properties.

Theoretical computations and experiments indicate that the tensile strength – the maximum stress (force per area) not causing rupture – is 50 to more than 100 GPa, which is 100-fold higher than for steel (0.3–1.8 GPa). With a density of 1.3 Mg/m^3 (1/6 that of steel, 7.9 Mg/m^3), carbon nanotubes show a record strength–density ratio. In other words, they promise much stronger material at reduced weight. Utilization will depend on success in producing longer fibres, spinning them into ropes, and forming composites. For the latter, the main problem is achieving high load transfer at the interface between nanotubes and matrix; adding functional groups for adhesion would probably weaken the nanotubes. If this problem could be solved, spectacular projects in principle can become possible, e.g. a so-called space tower or space elevator: a 100,000 km long cable fixed to an equatorial platform on earth, tensioned by the centrifugal force of the part extending

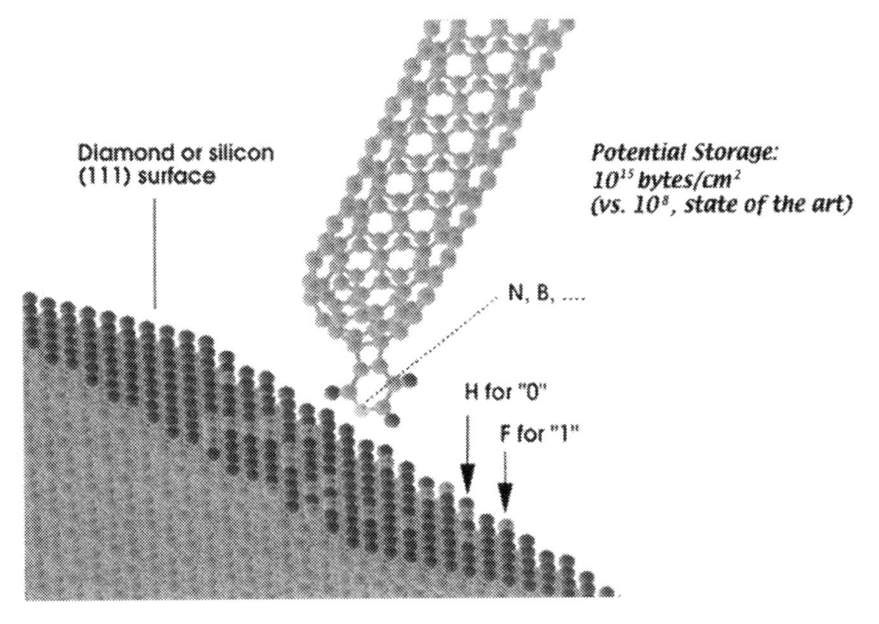

Figure 2.2 Carbon nanotube writing bits coded as hydrogen (H) or fluorine (F) atoms on a carbon or silicon surface in a NASA concept. (From NASA 2002a, reprinted by permission; original in colour.)

beyond the geostationary altitude of 36,000 km. It could carry a platform, deploy geosynchronous and other satellites and accelerate spacecraft to planets and deep space like a sling (e.g. Edwards 2000, 2000a).

Carbon nanotubes could be used for molecular electronics – first, as interconnects with very little resistance due to ballistic electron transport.[10] Second, one can form junctions with rectifying or switching behaviour. In a field-effect transistor, the current through a nanotube lying on top of an insulating silicon dioxide layer was controlled by the voltage at the silicon gate below. The conductivity of semiconducting nanotubes depends strongly on surface modifications and mechanical deformations. This provides a mechanism to use them in sensors for adsorbed gas molecules or for mechanical quantities such as vibration or pressure. Nanotubes have been attached to scanning-probe microscopes, resulting in particularly thin probes, similar to Figure 2.2.

Because they are so thin but can carry high currents, carbon nanotubes are well suited for field emission, releasing electrons at moderate voltages into a vacuum without the need for cathode heating. This can be used in flat displays or light sources.

Carbon nanotubes have been proposed for lithium storage in batteries; they may also be used for super capacitors. Despite initial hopes, it seems that hydrogen storage in nanotubes at room temperature is quite limited so that other forms of nanostructured carbon, which are also cheaper, are more promising (Frackowiak and Béguin 2002).

2.2 Molecular NT

2.2.1 Assemblers and molecular NT

Since the mid-1980s, the term NT has been used in a narrower sense to denote manipulation at the atomic and molecular level by special molecular machines, and in particular the visionary concept of a universal molecular assembler (Drexler 1981, 1986). More recently, this branch of NT thinking has been called 'molecular NT' (MNT) by its proponents.[11] The assembler concept and others are usually attributed to Drexler, but many of them have been discussed since at least the 1960s. These include molecular engineering, molecular computers, molecular machines, artificial nanoreplicators, artificial evolution, nano-robots for cell repair, life extension.[12] R. Feynman (1959), in his famous speech, talked about automatic production of extremely small parts by a billion small machines that were produced by successive stages of miniaturization;[13] mechanical surgeons small enough to enter blood vessels; inspiration by biological systems; synthesis of arbitrary chemical substances (chemical stability permitting) by manoeuvring atom by atom.

For MNT, the biochemical processes in the living cell and organism serve as the great model. These processes in fact take place on the molecular scale: information-carrying molecules (DNA) are being read, according to this code protein factories (ribosomes) take specific amino acids from the surrounding fluid and assemble them into proteins, while proteins work together to catalyse the synthesis of whatever else is needed. The MNT visions comprise, on the one hand, full understanding of the life processes, their repair and eventual manipulation. On the other hand, MNT is about applying general concepts found in life to artificial systems – constructing useful objects from building blocks on the molecular scale, using molecular information carriers and working with the little energy that is available locally. In such a way, systems should become possible that have been intentionally designed to fulfil a certain purpose, transcending by far the capabilities that have evolved by chance and selection over the billions of years of natural evolution.

A central concept of MNT is the universal molecular assembler – a program-controlled molecular machine that synthesizes arbitrary molecules and larger units by selectively taking existing building blocks (single atoms only rarely) from a feedstock or the environment, and mechanically moving them to the intended place with atomic precision where they form the intended bonds, with some energy input if required (mechanosynthesis) (Drexler 1981, 1986; Drexler and Peterson 1991; Drexler 1992). In particular, the assemblers could self-replicate, with their number growing exponentially (as long as limits of resources are not reached). High numbers are essential for production on a macroscopically relevant scale.[14] After the required number of generations, assembly of the intended end products would start, maybe via several intermediate steps of micro- and macro-size production machines/robots.

According to the molecular-NT concept, any arrangement of atoms that is not excluded by the laws of physics and chemistry can be achieved – not only traditional artefacts, but also living organisms, and many new types of objects that before existed neither in nature nor as human products. For most purposes, no expensive raw materials would be needed – light elements that can be found nearly everywhere would suffice, such as carbon, nitrogen, oxygen, calcium etc. As in biology, synthesis could be very energy-efficient, and could to a large degree rely on solar power. Different from biology, systems could work outside of aqueous solution and in a much wider range of environmental conditions – temperatures from arctic to fire, pressures from ocean bottom to outer space, caustic or toxic surroundings. Structural frames and surface layers could consist of the hardest and most robust materials, e.g. diamond. Goods production by assemblers would be very cheap and autonomous, needing human work (in theory) only at the directing level or at the beginning.

By continuing the self-replication process, the number of end products could grow exponentially, too – as long as limits of locally available resources and energy, or of transport from farther away, would allow. Exponentially growing production may not be essential for civilian uses, but it could become relevant in scenarios of military dominance and arms races (see Section 4.3).

If assemblers, nano-robots and self-replication were to arrive, they would of course strongly accelerate all the other developments. This arrival would probably not occur as the 'assembler breakthrough', a short time after which all principal possibilities would be achieved. An MNT proponent has already remarked that a 'two-weeks revolution' following construction of the very first assembler, based on earlier simulations, is unlikely. Errors in first designs will only show up in practice, and as for software, a tedious learning process will be needed to remove the 'bugs' (Kaehler 1996).[15] Acceleration is conceivable as design, simulation, construction and experimenting could be taken over by fast AI, but even under such circumstances growth of capabilities would need time, if only by the laws of physics.

One company has been set up with the aim of actually realizing the MNT vision: Zyvex at Richardson, Texas, USA, was founded in 1997 by James R. Von Ehr II, an entrepreneur from the software industry (Ashley 2001; Zyvex 2003). The firm of thirty-seven staff (2001) has good connections to the Foresight Institute; institute members (e.g. R. Merkle and R. Freitas) have moved to Zyvex, the 4th Foresight Conference is archived at the Zyvex web pages (Foresight 1995). To approach the goal of a nano-assembler, Zyvex has started to work on MST and nanomanipulators. The latter research tools are their first actual products to be offered (Zyvex 2003a).

2.2.2 Associated concepts

MNT is associated with several additional futuristic concepts, as shown in Table 2.2. Some of them are independent of molecular assemblers and self-replication, but these would support the others. Assuming that atomically precise molecular manipulators are at hand, it is logical to presume that they could also be used as probes, e.g. in living cells. This would, on the one hand, lead to full understanding of the biochemical processes. On the other hand, it would provide the capability to modify and manipulate the cellular processes, e.g. for stopping and re-starting them (biostasis). Medical nano-robots could repair damaged DNA, prevent or reverse ageing

Table 2.2 Futuristic concepts associated with MNT (see text)*

Mobile nano-robots
Universal molecular assembler Self-replication Larger autonomous production machines
Nanocomputers
(Super)human artificial intelligence Automatic construction, automatic research
Modified biochemistry, biostasis
Improved organs Artificial organs
Nano-robots in cells for action on DNA, protein synthesis, etc.
Eradication of illnesses, of ageing
Nano-robots in neurones to sense or control
Read-out of brain contents Evocation of sensory impressions, of thoughts
Downloading of brain contents to software
Continued existence in software
Brain implants to expand memory, thinking, feeling, for communication
Merging of humans and robots
Outer space: mining of asteroids, extraterrestrial colonies, interstellar travel

Notes
* For more concrete ideas for MNT applications inside and outside the body, see the contributions in Crandall 1996.

processes (Drexler 1986: Chs 7, 8; Freitas 1999; Haberzettl 2002). With the help of MNT, artificial or hybrid organs or organelles could be developed that are more efficient than the natural versions, e.g. so-called respirocytes (micrometre-size nanomechanical devices with pressure vessels, molecule pumps, nanocomputer etc.) could carry more than 200 times the oxygen amount per volume than red blood cells (Freitas 1996/99). Teeth and bones could be strengthened by diamond (Reifman 1996).

With the capability to build (nearly) arbitrary structures on the nanometre scale, extremely small data-storage and -processing structures could be built. Memory sizes and processing speeds would increase by many orders of magnitude. 'Genuine' artificial intelligence (AI) would arrive, either through the complexity of conventional information systems, or perhaps by mimicking neuronal processing and learning. A system similar to the human brain could be much smaller, yet millions of times faster than the latter. AI would reach human levels of competence in a few decades and then fast transcend it. Doing experiments and developing new systems using assemblers, 'thinking machines' would advance technology fast to the limits set by the laws of nature (Drexler 1986: Ch. 5; Moravec 1988: Ch. 2; Kurzweil 1999).[16]

Nano-robots in neurones could sense not only the state of the complete cell (firing or not), but also the situation at individual synapses. In such a way, some writers claim that a complete image of a brain could be formed, down/uploaded to a computer and run there as a simulation, or as a continuation of that person's thinking process. Vice versa, the nano-robots could take control of the neurones, creating arbitrary sensory impressions, transferring thoughts or making connections to internal or external memory or processing devices, or to other peoples' brains (Moravec 1988: Ch. 4; Kurzweil 1999: Ch. 7).

With cheap production and lighter as well as stronger materials, MNT would provide inexpensive access to outer space. Autonomous, self-replicating systems could mine the moon, planets and asteroids, solving the problem of limited resources on Earth (Drexler 1986: Chs 6, 8, 10; McKendree 1998 and refs). Alternatively, MNT would provide the potential for long-distance human space travel and space colonies; some MNT proponents see this as indispensable because of increasing environmental pollution on earth, because of a high probability of a catastrophic extinction – whether through collision with a large meteor, a nuclear war or an ecological catastrophe – or to accommodate the ever-growing number of humans if ageing and death are overcome (Moravec 1988: 101; MMSG 2000; NASA 2002; see also Bostrom 2002; Rees 2003).

Connections between MNT in the narrower sense and the wider futuristic ideas exist also in the writings of key authors, and in institutional/personal links. While Drexler described many of the wider concepts, Moravec mentioned atom-by-atom synthesis by molecular robots (Drexler 1986; Moravec 1988: 73). Kurzweil emphasized the importance of MNT and has recently joined the Board of Advisors of the Foresight Institute (Kurzweil 1999: Ch. 7; Foresight 2002). The prominent Foresight Institute author R. Freitas has joined the editorial board of the Journal of Evolution and Technology, published by the World Transhumanist Association (Jetpress 2003). In this context, it is interesting that W.S. Bainbridge, co-editor of the proceedings of the conferences of the US NNI on societal implications of NT and on NBIC convergence (Roco and Bainbridge 2001, 2003; see Section 2.3), has given a speech at the awards ceremony of the Transvision 2003 conference of the World Transhumanist Association, arguing fervently against limits on human cloning, for projects on 'uploading personalities to a computer' and 'infusing AI with humanity'.[17]

It has to be noted that several of the wider concepts associated with MNT by its proponents are relatively independent of NT. Biomedical or artificial-intelligence applications, for example, might materialize without/before universal molecular assemblers. Instead of a sudden jump to radically new possibilities there would rather be a sliding transition. On the other hand, several of the concepts may turn out infeasible for a very long time – or for good.

2.2.3 MNT and mainstream science

2.2.3.1 MNT and the laws of nature

Many of the concepts associated with MNT sound speculative, but most would be compatible with the laws of nature as they are presently known. In particular Drexler has stressed that MNT has its limits in the laws of physics. As an example, he states that miniaturization of usable structures has to stop at the level of complete atoms; atomic nuclei, which are about 10^5 times smaller than atoms (several fm (femtometre = 10^{-15} m) instead of several 0.1 nm), cannot be brought close together because of fierce electrostatic repulsion (Drexler 1986: Ch. 10).

In contrast, computer technologist (and non-physicist) R. Kurzweil has postulated continuing shrinking of computer structures, reaching the picometre (10^{-12} m) scale in 2072 and femtometres around 2112 with structures in sub-atomic particles (e.g. electrons) and in quarks, respectively (Kurzweil 1999: Ch. 12, Time Line).[18] As Drexler has already remarked, however, structures of such small sizes cannot remain stable under ordinary conditions due to electrostatic repulsion; a theoretically imaginable exception is ultradense matter as in neutron stars – handling of which by normal matter is difficult to conceive.[19] One may consider the possibility of electromagnetic or other interactions involving nucleons in ordinary matter, but I know of no plausible suggestions of a physical basis for such interactions to be exploited for a sub-nanometre computing technology – and this holds even more for structures inside the particles that make up nucleons, that is quarks.

Also when discussing brain analysis and brain-content exchange, some visionary computer scientists do not care to discuss problems of concrete implementation: Kurzweil states that some day one will be able to measure non-invasively all 10^{14} synapses of the 10^{11} brain neurones. His argument is that the resolution of brain imaging techniques such as optical imaging or nuclear magnetic resonance will steadily improve, together with the growth of computing power (Kurzweil 1999: Ch. 6). However, there may be physical limitations to that resolution if one has to measure from outside the head, or the time needed for sequential focusing on all volume elements may remain too long.

According to H. Moravec, the natural brain could be replaced by an improved artificial one in a stepwise process. Layer after layer, the chemistry and pulses of the neurones would be analysed and a simulation program started. When the awake 'patient' has tested (by pressing a button) that there is no difference between using the original cells and the program, the latter is connected permanently and the brain layer is removed for good (Moravec 1988: Ch. 4). Several questions are not addressed: how could one contact all the neurones without disturbing them? How about the connections to deeper-lying neurones? How would the architecture of neurones be analysed that did not fire during the surgery, because the patient did not have the thoughts or sensory impressions involving these neurones?

2.2.3.2 MNT and 'normal science'

It is remarkable that the mainstream science community has practically ignored MNT and related ideas.[20] Despite the potential importance of the topic, the finding of the former US Office of Technology Assessment (OTA) of 1991, that 'there has been little written criticism of molecular machine concepts' (OTA 1991: 21), still applies today. Whereas there are several articles on details of molecular assemblers in the scientific literature,[21] I am not aware of any refereed paper criticizing the basic concept. Some scientists working in or observing 'normal' NT have made short critical remarks within conference contributions or other texts (e.g. Tolles 2001; Vogel 2001; Harper 2003). The most systematic arguments that I have found are contained in two articles in *Scientific American* that obviously have not been written for peer review. R. Smalley maintains that 'self-replicating, mechanical nanobots are simply not possible in our world' (Smalley 2001). G. Whitesides, on the other hand, states that such systems would be rather biology-like; others would not arrive in the foreseeable future (Whitesides 2001). MNT proponents quickly published detailed refutations (Drexler *et al.* 2001, 2001a). A later exchange of opinions, however, did not bring much clarification (Drexler and Smalley 2003). OTA conjectured a 'reluctance of scientists to denounce new concepts in publications' (OTA 1991: 21).[22] Many mainstream scientists seem to be sceptical of MNT; on the other hand, several have accepted the Foresight Institute Feynman Prizes in NT.[23]

Concerning nanomachines as such, the leader of the group which first powered an inorganic nanodevice with a biomolecular motor wrote that the idea of 'complex, molecular-sized engineered devices' 'seamlessly interfacing ... with fundamental life processes' is moving from science fiction to an 'achievable goal' (Montemagno 2001).[24] In contrast to molecular assemblers, the concept of computers reaching and transcending human intelligence has not met universal scepticism in science – researchers in the fields of computing, robotics, or AI have often taken such developments for granted or even advertised them.[25] The US NT Research and Development Act of December 2003 requests the National Research Council to carry out, in the first three years, a study of 'the technical feasibility of molecular self-assembly for the manufacture of materials and devices', and a 'Study on the Responsible Development of Nanotechnology' covering self-replicating nano-machines, enhancement of human intelligence etc. (Congress 2003: Section 5).

The relationship between 'normal' science (scientists and their community, planning and funding institutions, etc.) and proponents of MNT is certainly an interesting field for the sociology of science.[26]

2.2.3.3 The approach to MNT in this book

The approach towards MNT in this book is guided by the precautionary principle: those concepts that do not obviously run counter to the laws of nature will be taken seriously as principal possibilities. Such concepts could only be safely ignored if it were reliably demonstrated that they are technologically impossible. Because the question is open as to whether universal assemblers, self-replicators, etc. are realizable at all, and if so, in which time frame, the discussion of MNT is necessarily more speculative, and will be presented in separate sections.

2.3 Convergence of nano-, bio-, information and cognitive sciences and technology

Some of the broader MNT-related futuristic concepts do not rely on universal assemblers, nano-robots or self-replication. In particular, progress in medicine, molecular biology/genetics/proteomics, electronics, robotics and software may lead to implanted brain interfaces, manipulation of genes and biochemistry in body cells, very small but extremely powerful computers, AI of human or greater ability. Developments in such areas will mutually accelerate each other, in particular because at the nanoscale NT, biotechnology, information technology and neuroscience converge.

This was acknowledged in the high level of US government research planning and funding by the remarkable workshop 'Converging

Technologies for Improving Human Performance – Nanotechnology, Biotechnology, Information Technology, and Cognitive Science' (NBIC) that took place in December 2001 in the USA at the request of the Interagency Subcommittee on Nanoscale Science, Engineering and Technology (NSET),[27] sponsored by the National Science Foundation (NSF) and the Department of Commerce (Roco and Bainbridge 2003).[28] In six groups, the workshop discussed technological progress possible from the confluence of these four areas of science and technology in the next ten to twenty years and recommended strong efforts towards these goals (Section 1.5.10 described the national-security group). In addition, it speculated about future prospects beyond twenty years.

Sidestepping universal molecular assemblers and self-replicating nanorobots, many of the related concepts which have been promoted by MNT visionaries were presented as realistic possibilities, including: full understanding of the human mind and brain, augmenting brain memory, NT-based implants as replacement for human organs, nano-robots for medical intervention in cells, slowing down or reversing ageing, brain-to-brain and brain-to-machine interfaces, robots and intelligent agents that embody aspects of human personality, computer-based social-science prediction of society and advanced corrective actions. The need for ethical considerations is mentioned, and the existence of risks is acknowledged – in general terms – at places. Unfortunately, no systematic discussion of risks and preventive measures is included.

2.4 Areas of NT

NT comprises very many areas and aspects. They can be subdivided according to various criteria, such as the degree of complexity of the structures, the closest scientific discipline, the production process, the envisioned application, the time scale of potential introduction.

With respect to degree of complexity, the areas of NT can be subdivided as in Table 2.3 where they are ordered according to increasing dimensionality.

NT is an interdisciplinary endeavour. In the analysis of phenomena as well as in the design of systems at the nanoscale, the borders between the different scientific disciplines become blurred – physics, chemistry, biology, medicine, computer science and their respective sub- and intermediate disciplines, such as mechanics, electronics, biochemistry, genetics, neurology, artificial intelligence and robotics, meet according to the respective object of study. This is reflected in the concept of convergence of nano-, bio-, information technology and cognitive science (see Section 2.3).

Table 2.4 shows examples of how major disciplines relate to NT, Table 2.5 gives exemplary production processes of NT.

Table 2.3 Subdivision of NT areas according to degree of complexity (dimensionality increases from point-like to three)

Object class	*Example*
Homogeneous/periodic (bulk)	
Powder of nanoparticles – also in solvent	Paint, sunscreen
Nanotubes, nanowires	Carbon nanotube
Simple layer with nm thickness, e.g. by adsorption of molecules from solution, often in preferential orientation by self organization; also by fixing of powder by firing	Diamond layer for hard surface, monolayer for molecular electronics, protective coatings, solar-power generation
Three-dimensional layered arrangement	Magnetic storage
Three-dimensional periodic/random arrangement	Protein crystal, zeolite, three-dimensional molecular memory
Complex structure	
Linear chain	Information-carrying molecule (as DNA)
Mostly on a surface, little depth	Scaled-down microelectronics, nanomechanical device
Produced by surface techniques, but many layers	Vertical-cavity surface-emitting laser (VCSEL),...
Fully three-dimensional, no self-replication	DNA scaffold, biomolecular computer, nanomachine
Fully three-dimensional, with self-replication	Self-replicating nano-robot

2.5 NT research and development

In the early 1990s, efforts for NT R&D were increasing slowly in the most-developed industrial countries. In the late 1990s, recognition grew that NT could provide radically new possibilities in many areas, and special programmes were consolidated or newly founded. International competition has played a strong role here.

In the *USA*, in 1996–1998 the world-wide status of and trends in nanostructure science and technology were assessed for several government agencies, led by the National Science Foundation.[29] This study compared the activities in Europe, Japan and the US in gross terms. For the six areas considered, the US and Japan led in two each, and in the remaining two Europe and the US were on a par (Table 2.6).

This process led to the founding of the National NT Initiative (NNI) in late 2000.[30] In the following years, US spending on NT R&D has been strongly increased. As a reaction, NT funding in Japan and Europe was increased, which was in turn used as an argument to raise funding in the USA.[31] Table 2.7 and Figure 2.3 show the funding increase in various regions

Table 2.4 Scientific/technical disciplines and example relations to NT – in many cases, more than one discipline is involved

Discipline/field	*Examples*
Physics	Tools for analysis and manipulation: scanning-probe microscopes, near-field optical microscope, optical tweezers Electronic/magnetic/optical properties of nanostructures Nanomechanics Self organization
Chemistry/Materials Science	Particles Coatings Porous material Dendritic molecules Nanofibre composites DNA-based scaffolds
Electronics	Optical/electron-/ion-beam lithography Layered magnetic sensor Mechanically moved probes for data storage Nanotubes as conductors, switches
Information Science	DNA computing Molecular computing
Biology	Analysis of biomolecules Analysis of cell processes Biomineralization Biological motors Biocomputer
Medicine	Particles coated with antigens/antibodies/DNA patterns: nanodots for optical signalling, magnetic for separation or heating; for ferrying therapeutic drugs across barriers DNA chip Biocompatible materials Implants for analysis and drug release Electrodes for nerve/brain contact

Table 2.5 Examples of NT production processes

Particles from gas phase (flame, plasma)
Sol-gel process for composites
Optical lithography, electron-/ion-/atom-beam lithography
Stamping, imprinting
Self-assembly
Scanning-probe microscopes: manipulation of individual atoms/molecules on a surface
Mechanosynthesis (molecular NT)

Table 2.6 Comparison of activities in nanostructure science and technology in Europe, Japan and the USA 1996–1998 (Siegel *et al.* 1999: xxi)

Activity	*Level*		
	1	*2*	*3*
Synthesis and assembly	USA	Europe	Japan
Biological approaches and applications	USA/Europe	Japan	
Dispersions and coatings	USA/Europe	Japan	
High surface area materials	USA	Europe	Japan
Nanodevices	Japan	Europe	USA
Consolidated materials	Japan	USA/Europe	

of the world. (For the distribution of the NNI funds among the various US agencies see Table 3.1 in Section 3.1.1.) In 2001, more than thirty countries had NT activities and plans (Roco 2001). Table 2.8 shows the fifteen countries with the highest shares of NT-related publications and patents.

In *Germany*, the Federal Ministry of Education and Research (BMBF) had funded NT research since the beginning of the 1990s in its programmes Material Research and Physical Technologies, later also in others. In 1998, it founded Competence Centres for more focused work and networking between all NT actors (Table 2.9). In the following years, the NT budget was continuously increased (Table 2.10). The total government funding for NT is higher: in 2001, the Federal Ministry of Economics and Technology gave about €6 million, and the NT share of the institutional funding for the German Research Foundation (DFG) and various research institutions (paid by BMBF and the Federal States) amounted to €93 million. All told, public NT funding in 2001 was €153 million and in 2002 about €198 million (BMBF 2002).

In the *European Union*, NT topics were covered under various headings in the 4th and 5th Framework Programmes for Research, Technological Devel-

Table 2.7 Estimated government expenditures for NT R&D in $ millions (Roco 2001, 2003). Uncertainties due to: exchange rate, earlier US fiscal-year start, definition of NT, difference between allocations and spending. Western Europe includes countries in EU and Switzerland, 'Other' includes Australia, Canada, China, Eastern Europe, Former Soviet Union, Israel, Korea, Singapore, Taiwan and other countries with NT R&D

Area	*1997*	*1998*	*1999*	*2000*	*2001*	*2002*	*2003*
Western Europe	126	151	179	200	~225	~400	~650
Japan	120	135	157	245	~465	~720	~800
USA	116	190	255	270	465	697	774
Other	70	83	96	110	~380	~550	~800
Total	432	559	687	825	1,535	2,367	3,024

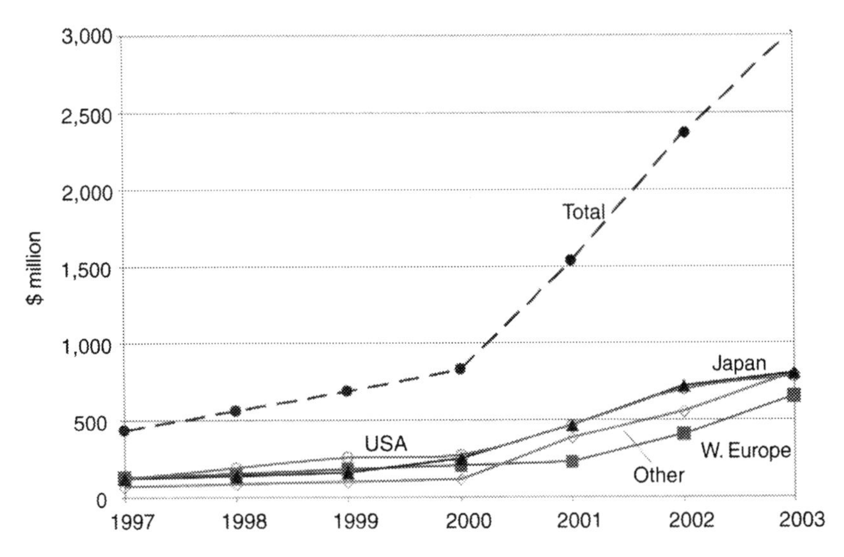

Figure 2.3 Estimated world-wide government expenditures for NT R&D (from Table 2.7).

Table 2.8 The fifteen countries with the highest shares in NT publications 1997–1999 (left) and in NT patents 1991–1999 (right), ordered by share (Companó and Hullmann 2002)

USA	USA
Japan	Germany
Germany	Japan
China	France
France	UK
UK	Switzerland
Russia	Canada
Italy	Belgium
Switzerland	Netherlands
Spain	Italy
Canada	Australia
South Korea	Israel
Netherlands	Russia
India	Sweden
Sweden	Spain

opment and Demonstration Activities (1994–1998, 1998–2002). In the 6th Framework Programme (2002–2006), a specific Thematic Area 'Nanotechnology and nanosciences, knowledge-based multifunctional materials, new production processes and devices' was created. The budget comprises €1.30 billion over four years (EU FP6 NMP 2003). Of this, 25–30 per cent are for NT; adding the NT funding in the other thematic priorities (life sciences, information sciences etc.), the NT-related expenses are about €700 million, corresponding to an annual average of €175 million (Roman 2002). The

Table 2.9 NT Competence Centres funded by the German Federal Ministry of Education and Research (BMBF) since 1998. In 2002, these centres had between fifty-three and 113 members each (BMBF 2002)

Lateral nanostructuring
Nano-optoelectronics
Nanochemistry
Ultrathin functional layers
Nanoanalytics
Ultraprecise surface machining
Materials of nanotechnology

Table 2.10 Funding of NT R&D by the German Federal Ministry of Education and Research (BMBF) in € million (2002 and 2003 plan) (BMBF 2002)

	1998	*1999*	*2000*	*2001*	*2002*	*2003*
Joint projects	27.0	31.1	32.7	52.0	86.7	110.6
Networking by competence centres	0.6	1.6	2.1	2.1	1.8	1.5
Total	27.6	32.7	34.8	54.1	88.5	112.1

European Nanobusiness Association cites certain EC programme officials who estimate that 30 per cent of the whole spending will be NT-related; this argument leads to an estimated €850 million per year (Roman 2002).[32]

Of course, publicly funded work represents only a part of the NT R&D effort. On the one hand, public money is often given on a cost-shared basis. For the European Union, for example, one estimate gave an additionally mobilized capital of around €700 million per year (Roman 2002). On the other hand, there are significant R&D efforts undertaken by enterprises on their own. Large multi-national corporations are actively engaged, and venture capital is flowing into NT start-up businesses.

2.6 Expected NT market

Governments and industry expect huge markets for NT-based products. The US NNI gave examples of the world-wide sales to be expected in 10–15 years (see Section 1.4 for some of the expected applications) (Roco and Bainbridge 2001: 3–4):

- nanostructured materials and processes: $340 billion/year,
- electronics: $300 billion/year,
- pharmaceuticals: $180 billion/year,
- chemical plants: $100 billion/year,
- transportation: aerospace alone $70 billion.

The total world-wide market would thus amount to more than $1 trillion per year.

3

MILITARY EFFORTS FOR NANOTECHNOLOGY

Military research and development (R&D) at the nanoscale have gone on at a moderate pace for about two decades, more or less in parallel to civilian work. As in other military areas, the USA made the greatest effort. In recent years, the USA has strongly increased its military R&D for nanotechnology (NT) – a decision that may or may not be followed by other countries. Because the USA sets a clear precedent and because military R&D is most transparent there, the US efforts will be described in some detail in Section 3.1. For other countries, much less information is available. Some other countries are treated in Section 3.2, and their efforts are compared to the US ones in Section 3.3.

3.1 USA

Department of Defense (DoD) efforts to reach the nanoscale began twenty years ago, in the area of microelectronics (Murday 1999). Already in the early 1980s, a DoD programme had been started to reduce structural dimensions to below a micrometre. When the first scanning-probe microscopes were developed, they became a major focus of military R&D in the late 1980s. In 1996, nanoscience was named as one of six strategic research areas for Defense.[1] In 1997, DoD agencies spent about $32 million for NT research (the US government total was $116 million) (Roco 1999).[2] In fiscal year (FY) 1999, DoD investment in nanoscience and nanotechnology amounted to $70 million, out of which $50 million were for nanoelectronics. The research and technology agencies of the services (Army Research Office ARO, Office of Naval Research ONR, Air Force Office of Scientific Research AFOSR), as well as the Defense Advanced Research Projects Agency (DARPA) were actively funding various areas of NT, and the laboratories of the services (Army Research Laboratory ARL, Naval Research Laboratory NRL, Air Force Research Laboratory AFRL) were heavily involved (Murday 1999).

3.1.1 Military funding in the NNI

When the National NT Initiative (NNI) was founded in 2000, the DoD got a major share from the beginning; the DoD amount grew with the general strong increase of NNI funding and was above 1/4 of the total, second only to the funding of the National Science Foundation (see Table 3.1 and Figure 3.1).

Whereas in the first year, basic research (US DoD R&D category 6.1) was by far the most dominating, in the following years its share was around 45 per cent while applied research (6.2) and advanced technology development (6.3) got around 55 per cent (see Table 3.2). The categories further down the line (6.4 Demonstration and Validation, 6.5 Engineering and Manufacturing Development and 6.7 Operational Systems Development) that are directed towards a specific new product or upgrade are not, or not yet, included (6.6 is Management Support) (see Moteff 1999).

One investment mode of the NNI is for 'Centers of Excellence' (see Table 3.3). The DoD has founded three such centres: the Institute for Soldier Nanotechnologies at MIT (see Section 3.1.6), the Institute for Nanoscience at the Naval Research Laboratory (see Section 3.1.3.1) and the Center for Nanoscience Innovation for Defense at UCSB

Table 3.1 Funding for the NNI and the share of major agencies in US$ million (2000 to 2002: actual, 2003: appropriated, 2004: request) (Roco 2003, NNI 2003: 5–8). The initiative started in fiscal year 2000. Other participating agencies – without NT R&D budgets – are Department of State, Department of Transportation, Department of Treasury, Food and Drug Administration and Intelligence Agencies

Agency	*FY 2000*	*FY 2001*	*FY 2002*	*FY 2003*	*FY 2004*
National Science Foundation	97	150	204	221	249
Department of Defense	70	125	224	243	222
Department of Energy	58	88	89	133	197
National Institutes of Health[a]	32	40	59	65	70
National Institute of Standards and Technology[b]	8	33	77	66	62
National Air and Space Administration	5	22	35	33	31
Environmental Protection Agency	–	6	6	5	5
Department of Homeland Security[c]	–	–	2	2	2
Department of Agriculture	–	1.5	0	1	10
Department of Justice	–	1.4	1	1	1
NNI total	270	465	697	770	849

Notes

a Department of Health and Human Services.

b Department of Commerce.

c Transportation Security Administration.

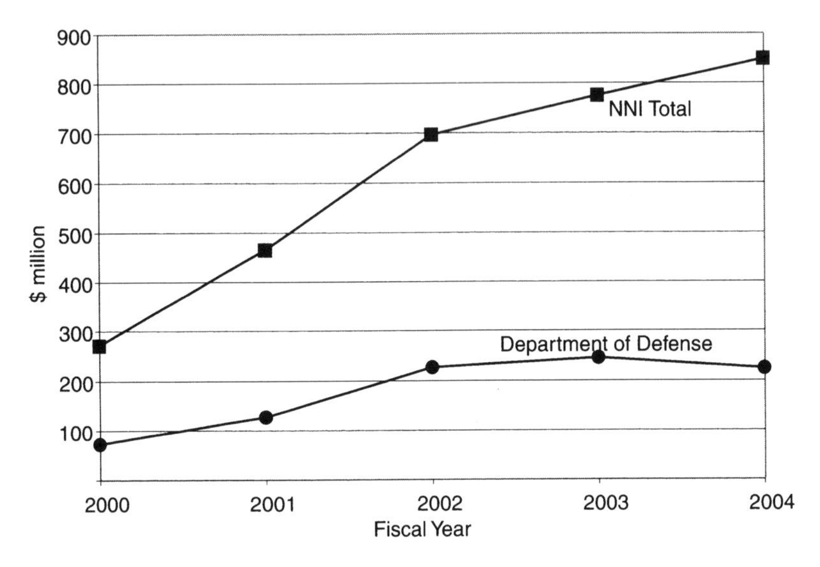

Figure 3.1 Funding for the NNI and share of Department of Defense (from Table 3.1).

Table 3.2 Breakdown of military funding in the US NNI for basic research (6.1), applied research (6.2) and advanced technology development (6.3), for the various DoD agencies, in US$ million (Roco 2002)

Agency	*FY 2001 (Actual)*		*FY 2002 (Plan)*		*FY 2003 (Request)*	
	6.1	*6.2/6.3*	*6.1*	*6.2/6.3*	*6.1*	*6.2/6.3*
DUSD (R)[a]	36	–	26	–	28	–
DARPA[b]	28	12	9	88	11	90
Army	6	–	18	2	18	5
Air Force	6	4	8	7	13	5
Navy	31	–	21	1	26	5
Total	107	16	82	98	96	105

Notes
a DUSD (R): Deputy Undersecretary of Defense for Research.
b DARPA: Defense Advanced Research Projects Agency.

(see Section 3.1.7) (on the civilian side, there are eleven centres – seven funded by NSF and four by NASA) (NNI 2003: 34).

It is interesting that in the 2004 Budget Supplement 'Intelligence Agencies' are mentioned for the first time. Their budget is not given and not included in the NNI sum. Their special interests in the respective investment modes are shown in Table 3.3. Looking at those, the ideas of small sensors and micro-robots for eavesdropping suggest themselves.

Table 3.3 Investment modes of the US NNI and interests of the intelligence agencies (NNI 2003)

Investment Mode	*Intelligence Agency Interests*
1 Fundamental Nanoscale Science and Engineering Research – Knowledge Generation	National security
2 NNI Grand Challenge Areas	
Nanostructured materials by design	Materials-by-design for intelligence applications
Manufacturing at the nanoscale	Prototype functional nanodevices
Chemical-biological-radiological-explosive (CBRE) detection and protection	Detection of CBRE agents
Nanoscale instrumentation and metrology	–
Nano-electronics, -photonics, and -magnetics	Molecular electronics and advanced communication systems
Healthcare, therapeutics, and diagnostics	–
Energy conversion and storage	Nano-enabled advanced power systems
Microcraft and robotics	Novel robotic systems
Nanoscale processes for environmental improvement	–
3 Centres of Excellence	–
4 Research Infrastructure	–
5 Societal Implications and Workforce Preparation	–

3.1.2 NT R&D funded by the Defense Advanced Research Projects Agency

3.1.2.1 Overview of programs and budgets

Within the US Department of Defense, DARPA gets by far the highest share of NNI funding, with a clear emphasis on applied research (6.2) and advanced technology development (6.3). Which DARPA programs come under the NNI heading could not be found easily, however. In order to get an overview of the NT-related work and an estimate for the corresponding expenditure, the DARPA Biennial Budget Estimates for Fiscal Years (FY) 2004–2005 were studied (DARPA Budget 2003). This 423-page document contains the RDT&E (research, development, test and evaluation) Budget Item Justification Sheets for the individual DARPA programs with short explanations. Excluding management, the 313 programs are grouped into three budget activities, thirteen program elements[3] and forty-eight projects.

For each program, the annual expenditures are given for FY 2002 through 2005. Summary figures for the program elements extend to FY 2009. Table 3.4 shows the program elements, their budgets for FY 2003, the number of projects of each and the respective number of programs.

In order to find whether a program is related to NT, all Budget Item Justification Sheets were examined. Programs that seemed to contain R&D in an area of NT proper were categorized as NT-related in a narrow sense. The second category comprises programs that deal with broader aspects of NT; this includes aspects of biology, artificial intelligence, cognitive science and robotics. Some of these may not yet use NT, but are likely to profit from it in the future, at least from smaller, more capable computers. This categorization is superficial and somewhat arbitrary, and may have overlooked a few NT-related programs. Table A1 in Appendix 2 shows the programs related to NT in a narrow and a broader sense with their budgets for FY 2003. The numbers of such programs are given in the last columns of Table 3.4. Because the NT content in generic programs, such as for structural materials or small satellites, could not be separated, the expenses are an overestimate.

NT-related programs were found in eleven of the thirteen program elements. Of the 313 DARPA programs, thirty-six were classified as related to NT in a narrow sense, and fifty-one in a broader sense. The former comprise expenses of US$468 million, the latter of US$547 million in FY 2003. This is 17 per cent and 20 per cent, respectively, of the total DARPA budget of US$2,690 million.

These figures are much higher than the one listed for DARPA ($101 million) under the NNI, see Tables 3.1 and 3.2. The reasons are probably 1) that the respective program expenses contain work beyond NT and/or 2) that not all NT-related DARPA R&D comes under the NNI heading.

To give an impression of the depth and width of the NT-related work funded by DARPA, several programs will be presented in the following sections.

3.1.2.2 Some DARPA programs narrowly related to NT

In *electronics/computing*, four programs look into Advanced Lithography to develop microelectronics (and other structures) with below 50nm feature size (DARPA Budget 2003: 277–279). Fundamental alternatives to traditional computers are being pursued in several directions. Two programs are investigating spin-dependent electronics, materials and devices, such as spin transistors and quantum-logic gates (Spin Dependent Materials and Devices, Spin Electronics, DARPA Budget 2003: 22–23). The Moletronics program is aimed at integrating molecules, nanotubes, nanowires etc. into scalable devices; in the Molecular Computing program, combinatorial logic functions and memory are to be implemented in molecular components and integrated to form a demonstration processor (sequential logic/finite-state machine) capable of interpreting a simple high-level language (DARPA Budget 2003: 231–232; DARPA Mole 2003). The new challenges for nanoscale interconnects will be tackled in the program Interfacing Nanoelectronics (DARPA Budget 2003: 213–214).

Much more specialized is the Nano Mechanical Array Signal Processors program that investigates arrays of up to 1024 mechanical nano-resonators for radio-frequency signal processing. Applications could be in wrist-watch-size, low-power UHF communicators or navigation (GPS) receivers (DARPA Budget 2003: 291–292). Another example is the Chip-Scale Atomic Clock where nano-resonators would be used together with photonic and microsystems-technology (MST) components. Using alkali atoms in very small cavities, the extreme accuracy of an atomic clock ($\pm 10^{-11}$ relative) would be packed into less than $1\,cm^3$ (DARPA CSAC 2003). Such a clock would allow denser communication channels and higher jam resistance even for small carriers.

Concerning *materials*, there are many activities. Within the Structural Materials and Devices program – which spans a very wide range – R&D is carried out for large-volume, low-cost synthesis and assembly of nanomaterials and nanotubes with controlled attributes (DARPA Budget 2003: 197–198). Under Functional Materials and Devices, work is being carried out for conducting polymers for analog processing, electroactive polymers for displays and muscle-like sensing and actuation for robots, high-density magnetic memory, microwave materials (ferrites, nanocomposite ferroelectrics, magnetodielectrics, negative-index materials), functional (conducting, piezoelectric etc.) fibres for electronic textiles (DARPA Budget 2003: 200–201).

Biology-related R&D has increased strongly in recent years. The Nanostructure in Biology program, for example, looks into nano-structured magnetic materials using nanomagnetics to understand and manipulate individual biomolecules and cells. Biocompatible, nanomagnetic tags, sensors and tweezers and a cantilever-based magnetic-resonance force microscope are to be built. With cantilevers, spectroscopy and imaging at atomic resolution is to be applied to molecules and nanostructures (DARPA BioMagnetICs 2003; DARPA MOSAIC 2003; DARPA Budget 2003: 10–12).

For routine analysis and design of integrated biological/chemical microsystems, the program Simulation of Bio-Molecular Microsystems aims at modelling and demonstration of molecular recognition, transduction into measurable electrical and mechanical signals using nanopores, micro-/nano-cantilevers and nanoparticles, and fluidic/molecular transport on the micro- and nano-scale (DARPA SIMBIOSYS 2003; DARPA Budget 2003: 7–8).

Table 3.4 Overview of the DARPA funding structure for the period 2002–2005, evaluated from the Budget Estimates for FY 2004/5 (DARPA Budget 2003). Excluding management support, three budget activities comprise thirteen program elements. Planned expenses for FY 2003 are shown in US$ million. Program elements consist of projects that contain programs; in each program, work is done by a number of contractors. (Program elements and projects finished before 2003 are not listed. Project and program counts do not include the Classified and Management Headquarters program elements.) The final columns give the approximate numbers of programs that are related to NT in a narrow and a broad sense and the respective funding in US$ million; these data derive from Table A1 in Appendix 2

Budget activity	*Program element*	*Expenses FY 2003 ($ million)*	*No. of projects*	*No. of programs*	*NT-related programs*			
					Narrow		*Broad*	
					Number	*Funding ($ million)*	*Number*	*Funding ($ million)*
Basic Research BA1	Defense Research Sciences	199	4	26	15	136	2	36
Applied Research BA2	Computing Systems and Communication Technology	409	9	58	0	0	16	153
	Embedded Software and Pervasive Computing	59	3	6	0	0	3	27
	Biological Warfare Defense	162	1	12	1	37	1	5
	Tactical Technology	170	6	40	0	0	8	27
	Materials and Electronics Technology	434	5	49	12	188	3	59

Advanced Technology Development BA3	Advanced Aerospace Systems	235	2	22	1	40	8	141
	Advanced Electronics Technology	159	5	36	7	68	2	13
	Command, Control and Communication Systems	117	3	23	0	0	3	4
	Sensor and Guidance Technology	217	4	19	0	0	0	0
	Marine Technology	36	1	4	0	0	0	0
	Land Warfare Technology	166	3	12	0	0	3	82
	Classified Programs	288	?	?	?	?	?	?
	Network-Centric Technology	0	2	6	0	0	2	0
	Management Headquarters	42	(1)	(1)	–	–	–	–
	Totals	2,690*	48	313	36	465	51	547

Note

*Corrected for rounding error.

The Bio Futures program focuses on computation based on biological materials and interfaces between electronics and biology. It will create 2-nm-diameter channels for parallel processing of biomolecules, microfluidic devices for trapping insect embryos and a multi-cantilever field-effect transistor for measuring single-cell physiology. Algorithms for analysis of neuronal spikes, cellular regulation and tissue differentiation in embryos will be developed (DARPA Budget 2003: 8–9).

Nano-structured material is being studied for use in a bio-inspired lens of variable refractive index and thus controllable field of view (program Bio-Optic Synthetic Systems, DARPA BOSS 2003). Biomolecular motors produce rotating or linear motion from chemical reactions on the nanoscale. The corresponding program is to study their properties and integrate them into laboratory devices. Hybrid biological/mechanical machines could actuate materials and fluids at scales from nano to macro; application could be for sorting, sensing and actuating (program Biomolecular Motors, DARPA Biomolecular 2003).

The Biological Adaptation, Assembly and Manufacturing program studies adaptation to harsh conditions by specific genes to improve the stability of living cells and tissues, including platelets and red blood cells, and to reduce metabolism after injury. Assembly and manufacturing of bone, shell, skin etc. by nanoscale biomolecular networks is investigated (DARPA Budget 2003: 9–10).

In the Biological Warfare Sensors program, R&D of a great variety of systems is carried out. Narrowly related to NT are miniature sampling systems with new antibodies and 'designer small molecules' to bind specific agents (anthrax bacteria, pox viruses, toxins), and a bacterial biochip for the fast identification of species without the need for the DNA polymerase chain reaction (DARPA Budget 2003: 133–137).

3.1.2.3 Some DARPA programs broadly related to NT

Programs broadly related to NT may use NT indirectly or in the future. Many will incorporate NT at least in the form of improved computers, but a significant number also via sensors, structural materials and/or mechanical actuators. The former holds for artificial intelligence, the latter also for robotics. Biological, in particular biomolecular, work will profit from tools for investigation and manipulation at the nanoscale.

A special case is the program Quantum Information Science and Technology. It deals with theory and hardware components for quantum logic, memory, computing and secure communication (DARPA QuIST 2003). NT can come in via, for example, semiconductor nanostructures for quantum bits (electron states), single-photon sources and detectors.

In the field of *artificial intelligence* and *cognition*, DARPA programs are taking on some fairly bold tasks. There is no explicit reference to NT, but they implicitly rely on continuing increases in computer performance. Explicit goals are, e.g.:

> automation systems wherein there is no 'human machine interface' per se, but where the collaboration is so 'normal' that the interactions are fundamentally like human-human interactions (involving anticipation, mixed-initiative interactions, dialogue, gesturing, etc.).
>
> (Program Augmented Cognition, DARPA AugCog 2003)

> software technologies needed to program the autonomous operation of singly autonomous, mobile robots in partially known, changing, and unpredictable environments. These autonomous systems will not have to rely on . . . a remote human operator nor depend on high-quality, real-time/near real-time data link connnectivity which often cannot be guaranteed.
>
> (Program Mobile Autonomous Robot Software, DARPA MARS 2003)

> the creation of a new class of computational systems – Cognitive Computing Systems. These novel computer-based systems will reason, learn, and respond intelligently to things that have not been previously programmed or encountered. This will be accomplished by creating unique and powerful new abilities for computers to perceive and understand the world, and to reason intelligently with the results of this kind of perception.
>
> (Project Knowledge Representation and Reasoning, DARPA Budget 2003: 29–30)

These tasks build on decades of work with significant advances, so that at least some success is probable. Whether 'the seamless integration of autonomous physical devices, computation software agents, and humans' that is foreseen for the 'next transformational revolution for military force development' (DARPA Budget 2003: 101) will be achieved, remains to be seen. Within 20 years, it cannot be excluded.

The area of *autonomous vehicles/robots* is covered by several programs. For Future Combat Systems, the Perception for Off-road Robotics program is developing and testing revolutionary perception systems (hardware and algorithms) under various terrain and weather conditions for uninhabited vehicles in combat, including collective action (DARPA Budget 2003: 355–361). The program Tactical Mobile Robotics aims at semi-autonomous robot teams for land forces (DARPA Budget 2003: 341–345). Autonomous Software for Learning Perception & Control is to program robots for navigation, learning of new tasks and adaptation to new environments (DARPA Budget 2003: 57–71). The program Unmanned Ground Combat Vehicle is developing and testing prototypes with improved endurance, obstacle negotiation and transportability (small size); wheels, tracks or walking/crawling may be used (DARPA Budget 2003: 394–397). Similar work is underway for uninhabited combat air vehicles, including rotorcraft, partly in co-operation with the services (DARPA Budget 2003: 239–247). Planning, assessment and control of distributed, autonomous combat forces such as uninhabited combat air vehicles is the subject of the program Mixed Initiative Control of Automa-Teams (DARPA Budget 2003: 58–73).

In the area of *small robots*, the Eyes-On program envisions an air-launched micro-unmanned air vehicle that provides real-time imagery to a fighter pilot for confirmation of targets, avoidance of collateral damage and bomb-damage assessment. Communication will be by line-of-sight radio-frequency link. By loitering in the target area, the system is also to be used for long-range weapons (DARPA Budget 2003: 188). For operations in urban exterior, underground and indoor environments, the Urban Robotic Surveillance System program will develop sensor systems and ground and air platforms, including communication routers and resupply of fuel or power. Small robots are not explicitly mentioned, but the missions mentioned (route clearing, flank protection, tunnel clearing, scout and peacekeeping operations) make clear that they are part of the task (DARPA Budget 2003: 189). Software technologies for large groups of extremely small micro-robots that act in co-ordination are developed in the program Common Software for Autonomous Robotics. A human operator is to communicate with and control the swarm as a whole (DARPA Budget 2003: 121–122).

For military uses of *outer space*, the Space Assembly and Manufacture program aims at very large, lightweight space structures. Micro-satellites for analysis of resources on non-terrestrial objects, miniaturized robotics for processing materials and building structures, propellants and power generation will be investigated (DARPA Budget 2003: 264).

In the *biology* area, the program Controlled Biological and Biomimetic Systems is devoted to understanding and controlling the basic functions of organisms. One- and two-way interfaces and communications with animals and 'animats' (artificial animals) will be explored. Projects come under the headlines of Vivisystems, Hybrid Biosystems and Biomimetics (DARPA CBS 2003). The first is about investigating insects and using them as sentinels for chemical or biological agents. In the second area, one project is on micro-electrodes in the brain of a monkey to derive motor signals and control a robot arm; another project uses electrodes in the rat brain to control the motion of the animal (see below and Section 4.1.17). The third group covers, among others, flight stabilization, artificial muscles and biomimetic robots moving under water, climbing like a gecko and flying like an insect.[4]

The idea of the Engineered Tissue Constructs program is to grow a three-dimensional human immune system from stem cells *ex vivo*, including interactive engineering of organs. It would be used to test vaccines and immunoregulators (DARPA ETC 2003).

A few programs target the soldier's body. In order to have it adapt faster to extreme environments (temperatures, high altitudes etc.) and to increase survival after injury, research is being done on Metabolic Engineering for Cellular Stasis. A major focus is on long-duration preservation of blood and stem cells at reduced weight, to be reactivated on introduction into the body (DARPA MetaEng 2003; DARPA Fact 2003).

One goal of enhancing the human war-fighting efficiency is to prevent the effects of sleep deprivation. The program Continuous Assisted Performance aims at maintaining a high level of cognitive and physical performance over seven days, twenty-four hours each. To achieve this goal, methods from neuroscience, psychology, cell signalling and regulation, non-invasive imaging technologies and modelling will be used; among the means envisaged are magnetic brain stimulation and novel pharmacological approaches (DARPA CAP 2003; DARPA Fact 2003).

Another goal in providing 'superior physiological qualities to the warfighter' is to control energy storage and release in order to achieve, for example, 'continuous peak physical performance and cognitive function for three to five days, twenty-four hours per day, without the need for calories'. The Metabolic Dominance program will look at manipulations of metabolism, control of body temperature and ways of rapidly increasing the numbers and efficiency of muscle fibres and mitochondria (DARPA Metabolic 2003).

The Brain Machine Interface program aims at recording and understanding the neural excitation patterns in the brain connected to motor or sensory activity (Rudolph 2001; DARPA BMI 2003). The motor signals could be read and used to control a system directly, without the 'detour' via the efferent nerves and the muscles in, for example, arm and hand. Thus, triggering a weapon or manoeuvring an aircraft could occur a few tenths of a second faster. For closed-loop control, an appropriate form of sensory (visual, postural, acoustic, other) feedback is to be developed. In experiments with about 100 microelectrodes in the motor cortex of rats and monkeys, the intended motion profile could be derived and a robot arm controlled successfully in one and three dimensions (Wessberg *et al.* 2000; Nicolelis 2001; Nicolelis and Chapin 2001; see also Hoag 2003). The capability to read sensory signals in the brain could also be used for monitoring and communication. For human experiments and applications of reading or influencing brain patterns, non-invasive methods are envisaged for the time being.[5] Whether external sensors or stimulators can provide the required spatial and temporal resolution remains to be seen, however. As mentioned, the research is intimately linked with invasive animal experiments.

3.1.3 NT R&D at military research laboratories

3.1.3.1 Naval Research Laboratory

Research at the nanoscale is a long-term focus of the Naval Research Laboratory (NRL) at Washington DC. Among the services' laboratories, it used to have the largest share of NT-related work (Murday 1999). In 2001, NRL founded a special Institute for Nanoscience,[6] but work is also continuing in the Chemistry, Optics and Electronics Divisions. Table 3.5 gives a list of the research topics. James Murday, long-time Superintendent of the Chemistry Division, has been involved on behalf of the DoD in the NNI from its conception.[7] The NRL maintains a list of contacts for the nanoscience and NT work in the DoD laboratories and the respective funding agencies (NRL NT Labs 2002; NRL NT Funding 2002).

3.1.3.2 Army Research Laboratory

The US Army Research Laboratory (ARL) is doing 'aggressive' Nanomaterials Research in several areas (ARL 2002; see also Rudd and Shaw 2001). Table 3.6 shows a list of research directions. The *Nanotechnology for Chemical and Biological Defense* program aims at chemical and biological detection, decontamination and individual and collective protection systems. Three projects have already been concluded successfully and transitioned to potential users:

- nanoreactor-based reactive topical skin protectant,
- dendrimer-based handheld immunochromatographic assay for improved point detection of biological agents, and
- molecularly imprinted polymer-based chemical-agent sensors.

Development and testing of mild, non-toxic decontaminants for biological and chemical agents is underway. Another area is ultralight protective clothing systems that use thin layers of enzyme-based nanocapsules, polymer-based nanoreactors, nanoparticulates and perm-selective membranes for fuel cells and protective clothing.

The *Nanoscience for Structural Materials* program wants to use nanostructured systems 'in the design of ultralight material components with mechanical, thermal, barrier, and ballistic performance far superior to current capabilities'. In polymer materials, research focuses on the control of nanophase segregation and the dispersion of nanoparticulates into polymer matrices. Successfully developed were:

- techniques for fabrication, bonding and repair of fibre-reinforced composite using induction heating by dispersed magnetic nanoparticles,

Table 3.5 Research topics in the Naval Research Laboratory. Overlap may be due to joint projects between the Institute for Nanoscience and the traditional NRL Divisions (NRL Nanoscience 2003a; Houser and McGill 2002; Colton 2003; Kafafi 2003; Snow 2003; see also NRL NT Labs 2002)

Institute for Nanoscience	
Nanoassembly	Nanofilaments: Interactions, Manipulation and Assembly Chemical Assembly of Multifunctional Elements Directed Self-Assembly of Biologically-Based Nanostructures Template-Directed Molecular Imprinting Chemical Templates for Nanocluster Assembly Assembly of Laterally Coupled Molecular Nanostructures
Nano-optics	Photonic Bandgap Materials Organic and Biological Conjugated Luminescent Quantum Dots Organic Light Emitting Materials and Devices Nanoscale-Enhanced Processes in Quantum Dot Structures Nano-Engineered Photovoltaic Devices
Nanochemistry	Functionalized Dendrimeric Materials Polymers and Supramolecules for Devices
Nanoelectronics	Coherence, Correlation and Control in Nanostructures Neural-Electronic Interfaces Nanocluster Electronics by Macromolecular Templating
Nanomechanics	Nanoscale Measurement Techniques Nanochemical Resonators and Advanced Nanodynamics
Chemistry Division	
Material Chemistry	Gold Nanocluster Chemical Sensors Self-assembly/Characterization of Nanocluster Architectures Spatially-controlled Organic Chemistry on Semiconductor Surfaces Metal Nanoparticle Composite Materials Carbon Nanostructures (Theory)
Surface Chemistry	Nanostructured Mesoporous Materials/Aerogels Nanofilaments Carbon Nanotube Field Emitter Arrays Nanocrystal Synthesis Nanomechanics/Nanotribology High-index Silicon and Germanium Surfaces III-V Heterostructures/Cross Sectional STM Dip Pen Nanolithography InAs Nanostructure Sensor Single Molecule Biosensors

continued

Table 3.5 Continued

Optics Division	
Organic Light-Emitting Materials and Devices	2D Photonic Crystals/Photonic Band Gap Materials NSOM Studies of PBG Properties PBG Optoelectronic Components Future PBG Optoelectronic Devices based on GaN
Neural-Electronic Interfaces	Nanochannel Glass Microelectrode Arrays 2-D Multiplexer Arrays A Model Neural-Electronic Interface: Artificial Retinal Stimulation Neural Computation in Biological Systems
Biologically Conjugated Luminescent Quantum Dots	Luminescent Nanocrystalline Particles Electrostatically Driven Self-Assembly of Quantum Dot Bioconjugates Maltose binding protein (MBP): Fluorescence assay for maltose Detection of Trace Levels of Explosive (RDX) in Seawater
Electronics Division	
	Record Performance Nanometer-Scale InAs HEMTs Nanometer Scale InAs/AlSb Quantum Structures Gold Nanocluster Electronics and Sensors Semiconductor Quantum Dots: Growth and Spectroscopy Spin Dynamics in Semiconductors Theory of Semiconductor Nanostructures Nanofabrication: Breaking the 10 nm Barrier

- nanoparticle-reinforced composite coatings for improved ballistic performance of transparent soldier faceshields, and
- novel processing methods for polymer nanofibre membranes.

The *Nanoparticulate Materials* program investigates inorganic nanoparticulates and structures made from them. The main attention is given to boron carbide and tungsten. Remarkable is the development of the Nanogen, a microwave-plasma device which produces nanoparticles of several nm to several tens of nm size at rates of nearly 100 grams per hour. Milling is also being pursued and gives larger sizes. Sintering to full density and ballistic testing was done. A new process of plasma pressure compaction was developed for keeping the grain size during consolidation. Microwave sintering is being used, too. Among the goals are armour for personnel and light vehicles as well as refractory metals for anti-armour projectiles.

In *Nanoenergetic Materials*, ARL wants to develop new insensitive high-energy propellants for munitions with improved burning rate and mechanical properties. Nanocomposite reinforcement of new solid propellants can strengthen the molecular structure by intercalation. Such material can be designed for plasma-ignition technologies used in new electro-thermal gun systems. Other applications could be as rocket propellants.

The European Research Office of the US Army has sponsored two workshops in the United Kingdom on nanostructures in polymers (Rudd and Shaw 2001; Shaw 2002). Clay-derived silicates and carbon nanotubes may lead to marked improvements in elastic modulus and compression strength, electrical conductivity, flammability and thermal stability. Several research projects have begun in the UK.

Beside its own research, ARL is also involved in the Institute for Soldier Nanotechnologies (see Section 3.1.6).

3.1.3.3 Air Force Research Laboratory

The US Air Force Research Laboratory (AFRL) is also active in a variety of areas of NT (Busbee 2002; see also AFOSR 2002, 2003). Table 3.7 gives a list of research directions. In the Materials and Manufacturing Directorate, a NanoScience and Technology program has been founded. NT-based materials, electronics, sensors etc. are also relevant for the Air Vehicles, Space Vehicles and Sensors Directorates. Work is being done on nano-energetic particles for explosives and propulsion (Pomrenke 2002: 24).

Table 3.6 NT research directions in the ARL; keywords mentioned in an overview list of the DoD laboratories (NRL NT Labs 2002)

Biology	Nanobiodetection
Electronics	Quantum well IR sensors Semiconductor nanostructures Nanoelectronic materials Quantum dot VCSELs, quantum optics
Materials	Particulate nanocomposites Polymer nanostructures/nanomaterials Nanoceramics Modifications of energetic materials Nanostructured polymer network composites Growth of nanomaterials Nanomaterials for energetics and armour Nanostructured and nanopatterned materials
Physics	Quantum information science Nanophysics Mesophysics

Table 3.7 NT research directions in the AFRL; keywords mentioned in an overview list of the DoD laboratories (NRL NT Labs 2002)

Biology	Nanobiomimetics
Chemistry	Self assembly Nanoscale energetic materials
Electronics	Nanoelectronics Magnetic nanoparticles Molecular computing, nanostructure theory Quantum dots for hyperspectral devices Matter-wave nanolithography Terahertz
Materials	Nanostructures for high performance composites Multiscale computer simulations Carbon-based nanotubes/foams Carbon-based nanotubes and nanofibres Nanocomposites Nanophase metal and ceramics Nanostructured materials, nanocomposites, polymeric nanofabrication Materials for nanoscale device structures
Physics	Nanostructured optical materials Nanostructured optical materials; theory, modelling and simulation of nanomaterials Nanotribology Quantum computing and communications

3.1.4 Military NT R&D at national weapons laboratories

The three laboratories responsible for nuclear-weapons R&D – Los Alamos National Laboratory (LANL), Lawrence Livermore National Laboratory (LLNL) and Sandia National Laboratories (SNL) – have done NT-related work in the course of their usual activities. For stronger and more focused activities, SNL and LANL jointly founded the Center for Integrated Nanotechnologies in 2002 (CINT 2003; for Sandia work, see SNL 2003). This is one of five Nanoscale Science Research Centers funded by the Department of Energy in the NNI.[8] Still being built up, the Center lists as its research themes: nano-bio-micro-interfaces, nanophotonics and nanoelectronics, complex functional nanomaterials, nanomechanics, theory and simulation. Much of this work seems to be general research not directed to specific military applications.

At LLNL, NT R&D was strengthened and co-ordinated following the founding of the NNI. In the Chemistry and Materials Science Directorate, the Materials Research Institute has one of its two foci in Nanoscience and Nanotechnology.[9] In the same Directorate, there is a BioSecurity and Nanosciences Laboratory (CMS 2003: 15). Many NT-related projects are funded by the Laboratory-Directed R&D Program; some of these are listed in Table 3.8. Also here one gets the impression that a wide research area is being covered.

One example of specific military relevance is work on new nanostructured high explosives using aerogel technology. Using variable composition, the energy release can be programmed (Parker 2000a; CMS 2002: 41; see also Section 4.1.8). It is interesting that such R&D for new high explosives is also done under the Stockpile Stewardship Management Program that obviously is not only to make sure that existing nuclear weapons remain functional, but is tasked to 'enhance US defense capabilities through innovative materials and chemical R&D' (CMS 2002: 48). Computer modelling is used to investigate nitrogen fullerenes such as $C_{48}N_{12}$ that promise high explosives of higher energy density (Rennie 2003). One can assume that the three laboratories have additional secret programs on military NT uses, not only in the field of nuclear weapons.

Table 3.8 Selected projects to do with NT funded by the Laboratory-Directed R&D Program at the Lawrence Livermore National Laboratory (the total number of such projects is about thirty) (LDRD 2003)

Advanced sensors and instrumentation	Single-particle detection for genomes-to-life applications Carbon-nanotube permeable membranes Probing the properties of cells and cell surfaces with the atomic force microscope Photoluminescent silica sol-gel nanostructured materials for molecular recognition
Chemistry	Laser initiated nanoscale molecularly imprinted polymers
Engineering and manufacturing process	Nanoscale fabrication of mesoscale objects Nanofilters for metal extraction
Materials science and technology	Nanoscience and nanotechnology in nonproliferation applications Enhancement of strength and ductility in bulk nanocrystalline materials Dip-pen nanolithography for controlled protein deposition The properties of actinide nanostructures Hydrogen storage in carbon nanotubes at high pressures Exchange-coupling in magnetic nanoparticles composites to enhance magnetostrictive properties
Mathematics and computing sciences	First principles molecular dynamics for terascale computers Atomically controlled artificial and biological nanostructures (also in physics)
Physics	Smart nanostructures from computer simulation Colliding nanometer beams

3.1.5 Defense University Research Initiative on NT

For basic NT research to be carried out at universities, the DoD has introduced the Defense University Research Initiative on NT (DURINT) which is one of several University Research Initiatives and is part of the NNI (DURINT 2001). The DURINT programme is administered through ARO, ONR, AFOSR and DARPA. Proposals were sought for purchase of equipment supporting NT research and for fifteen specifically named research topics. For the research awards, a typical amount would be $0.5 to one million per year over three to five years. In 2001, seventeen equipment grants (total $7.25 million, Table 3.9) and 16 research grants (total $8.25 million in FY 2001, Table 3.10) were given; starting in FY 2002, up to $15 million per year will be spent (DURINT 2001a). In the context of the traditional Multidisciplinary University Research Initiative (MURI) of the DoD, further five projects were granted (Roco 2002). In addition, DURINT fellowships were awarded to fifteen scientists at eight universities (DoD fellowships 2001).

3.1.6 Institute for Soldier Nanotechnologies

In order to get NT-enabled systems closer to actual military use, the US Army is funding the Institute for Soldier Nanotechnologies. This effort dates back to at least July 1998, when several Army R&D institutions with NSF sponsored a conference and workshop.[10] An overview speaks of revolutionary advances in research, endless implications for the future soldier, self-replication capabilities on the horizon. Among the technology thrust areas discussed were materials, food, sensors, displays, power sources. In February 2001, ARO sponsored a Workshop on Nanoscience for the Soldier; much material has been published on the Internet (ARO Nanoscience 2001 and its links to workshop documents). Four working groups discussed requirements for need- as well as opportunity-driven research and critical proof-of-concept demonstrations for various areas (Table 3.11).

Some of the concepts are fairly wide-ranging, e.g. for materials: 'Make soldier invisible across the EM spectrum' (ARO Materials 2001). With respect to the soldier, the workshop discussed:

> Enhance muscle performance over current human performance
>
> . . .
>
> **Objective 7: Internal Data, Chemical, Communications and Signal (artificial systems within the soldier):**
>
> . . . This is a high risk, visionary program to develop internal measuring, monitoring, data processing and communications capabilities. . . .

Table 3.9 Equipment grants given to universities in the DURINT program in 2001 (DURINT 2001b)

Institution	*Title*	*Agency*
Arizona State University	Electron-Beam Lithographic Equipment	Navy
Harvard University	Scanning Probe Microscope for Microstructure Analysis	DARPA/ Navy
Kansas State University	III-Nitride Micro- and Nano-Structures and Devices – Growth, Fabrication, and Characterization	Navy
Lehigh University	A Focused-Ion Beam (FIB) Nano-Fabrication and Characterization Facility	Navy
Massachusetts Institute of Technology	A Comprehensive Experimental Facility for Nano- and Meso-Scale Mechanical Behaviour of Nano-Structured Materials and Coatings	Navy
Massachusetts Institute of Technology	Very Low Temperature Measurement System for Quantum Computation with Superconductors	Air Force
Pennsylvania State University	Photoelectron Spectrometer and Cluster Source for the Production and Analysis of Cluster Assembled Nanoscale Materials	Air Force
Rice University	Magnetic/RF Field Assisted Spinning of Carbon Single Walled Nanotube Fibres	Navy
Stevens Institute of Technology	The Science and Technology of Computational Nano/Molecular Electronics – Equipment Proposal	Army
University of California at Santa Barbara	Catalysis by Nanostructure: Methane, Ethylene Oxide, and Propylene Oxide Synthesis on Ag, Cu, or Au Nanoclusters	Air Force
University of Illinois at Urbana Champaign	Ultrafast Vibrational Spectrometer for Engineered Nanometric Energetic Materials	Army
University of Arizona	Nanotechnology Instrumentation	Air Force
University of Colorado	Equipment for Molecular Rotors	Army
University of Michigan	Science and Technology of Nanostructures for Advanced Devices	DARPA/ Navy
University of Texas	Acquisition of an Electron Beam Lithography System for the Study of Nanostructures Formed using Bioinspired Self-Assembly Processes	Army
University of Virginia	Acquisition of a High-Resolution Field Emission Electron Microscope for Nanoscale Materials Research and Development	Air Force
Western Kentucky University	Acquisition of an x-Ray Diffractometer for Nanotechnology Research	Air Force

Table 3.10 Research grants given to universities in the DURINT program in 2001 (DURINT 2001c; see also Pomrenke 2002)

Prime institution	*DURINT topic*	*Agency*
University of Colorado	Nanoscale Machines and Motors	Army
University of Washington	Molecular Control of Nanoelectronic and Nanomagnetic Structure Formation	Army
University of Minnesota	Nano-Energetic Systems	Army
Stevens Institute of Technology	Characterization of Nanoscale Elements, Devices, and Systems	Army
Rice University	Synthesis, Purification, and Functionalization of Carbon Nanotubes	Navy
Princeton University	Nanoscale Electronics and Architectures	Navy
Carnegie Mellon University	Nanoporous Semiconductors – Matrices, Substrates, and Templates	Navy
Massachusetts Institute of Technology	Deformation, Fatigue, and Fracture of Interfacial Materials	Navy
University of California at Santa Barbara	Nanostructures for Catalysis	Air Force
Massachussetts Institute of Technology	Polymeric Nanocomposites	Air Force
State University of New York at Buffalo	Polymeric Nanophotonics and Nanoelectronics	Air Force
Massachussetts Institute of Technology	Quantum Computing with Quantum Devices	Air Force
State University of New York at Stony Brook	Quantum Computing with Quantum Devices	Air Force
Northwestern University	Molecular Recognition and Signal Transduction in Bio-Molecular Systems	DARPA/ Air Force
University of Southern California	Synthesis and Modification of Nanostructure Surfaces	DARPA/ Air Force
Harvard University	Magnetic Nanoparticles for Application in Biotechnology	DARPA/ Navy

Table 3.11 Areas discussed at the 2001 Workshop on Nanoscience for the Soldier (ARO Materials 2001, ARO Power 2001, ARO Soldier 2001, ARO Displays 2001). For Soldier Status, time scales were given: near-term (N), long-term (L), visionary (V)

Working group	*Areas*	
Materials and Fabrics	High-strength, ultra-lightweight materials	
	Adaptive, multi-functional materials	
Power, Energy Distribution and Cooling	Fuel cells, batteries, microturbines	
	Backup power systems	
	Body cooling	
Soldier Status Monitoring and Modelling	Clinical laboratory and pharmacy on the soldier	L-V
	Soldier status direct monitoring	N-L
	Local area monitoring	N-L
	Soldier performance prediction and virtual prototyping	N-L
	Active water reclamation	L-V
	Total sensory and mechanical enhancement (biological)	L-V
	Internal data, chemical, communications and signal processing (artificial systems within the soldier)	V
	Integrated bioenergy devices for driving sensors	L-V
Displays, Detectors and Antennas	Displays	
	Antennas	
	Detectors	

> RESEARCH REQUIREMENTS:...
> All components (implanted and native) must communicate, be transduced and link to command center
> Develop pin-size, biocompatible nano computers that can transduce data from sensors, process and communicate with command
> Onboard integration and control of sensing elements, without external manipulation
> Biological input/output structures
> Determine where placement of automatic monitoring/response structures within the body should be.
> ...
> Use of metabolites and chemicals to drive soldier systems
> (ARO Soldier 2001)

Based on this and other workshops, the Army Research Office put out a solicitation in October 2001, leaving out the long-term and visionary concepts. The Institute for Soldier Nanotechnologies (ISN), for which universities could compete, was to focus on the broad categories of Soldier Protection and Materials Development and Protection.

> The Institute will be chartered to conduct unclassified basic research into the creation and utilization of materials, devices, and systems through the control of matter on the nanometer-length scale and into the ability to engineer matter at the level of atoms, molecules, and supramolecular structures. The Institute will also research techniques for generating larger structures with fundamentally new molecular organizations exhibiting novel physical, chemical, and biological properties and phenomena.
> (ARO Solicitation 2001: Section 3.1.2)

The ISN should 'serve as the Army's focal point for basic research into nanotechnology for application to the future soldier', and should perform co-operative research with industry and the Army R&D institutions. 'A large centralized research facility is envisioned which will house world-class scientists and an exceptional research infrastructure.' An outreach plan should specify transfer of research to the Army, including personnel exchange with Army laboratories, and interaction with industry, other military agencies and universities (ARO Solicitation 2001: Sections 3.1.2, 3.1.3.4).

In March 2002, the Army selected the Massachusetts Institute of Technology (MIT) for the ISN. The five-year contract contains $50 million, and industry will contribute an additional $30 million (MIT News 2002; Army News 2002; Talbot 2002).[11] Edwin Thomas, Professor in Materials Science and Engineering, was named as the Director. The founding industry partners are: Raytheon, Dupont and the Center for Integration of Medicine and Innovative Technology (where two hospitals and MIT are members). The staff will number up to 150, including thirty-five MIT professors from nine departments. The basic research will be unclassified, and its results will be published in the open literature; sensitivities with respect to foreign students will be solved on a case-by-case basis (ISN Q&A 2002). Applied research and development where industry or military secrets have to be kept will probably be carried out in industry, at Army institutions, or at the MIT Lincoln Laboratory.

The ISN is to 'dramatically improve the survivability of individual soldiers through nanotechnology research in three key thrust areas: protection, performance enhancement, and injury intervention and cure' (ISN 2002). For this purpose, ISN will focus on six key soldier capabilities; the work will be carried out by seven multidisciplinary research teams (Table 3.12).

A guiding vision is a battle suit that protects against bullets and chemical/biological warfare (CBW) agents, has strength to apply force for lifting heavy loads or to stiffen around wounds, and senses body state and CBW agents (Figure 3.2). Optically variable material is to change colour for adaptive camouflage and form reflective patterns at invisible-light

Table 3.12 Key soldier capabilities and multidisciplinary research teams of the ISN (ISN Research 2002 and its links; Mullins 2002)

Key soldier capabilities

Strong, lightweight structural materials for soldier systems and system components

Adaptive, multifunctional materials for soldier systems and system components

Novel detection and protection schemes for biological/chemical warfare threats and identification of friend or foe

Remote and local soldier monitoring systems

Remote and local, wound and injury triage and emergency treatment systems to enhance soldier survivability

Novel, non-combat and combat performance enhancement systems for the soldier system that would improve soldier survivability en-route to and in the battlespace

Multidisciplinary research teams	*Funding in FY 2002 (US$ million)*
1 Energy Absorbing Materials	2.45
2 Mechanically Active Materials and Devices	2.25
3 Sensors and Chemical and Biological Protection	2.75
4 Biomaterials and NanoDevices for Soldier Medical Technology	1.35
5 Processing and Characterization	2.70
6 Modelling and Simulation	1.38
7 Technology Transitioning – Research, Outreach, Teaming with Industry and the Army	1.44

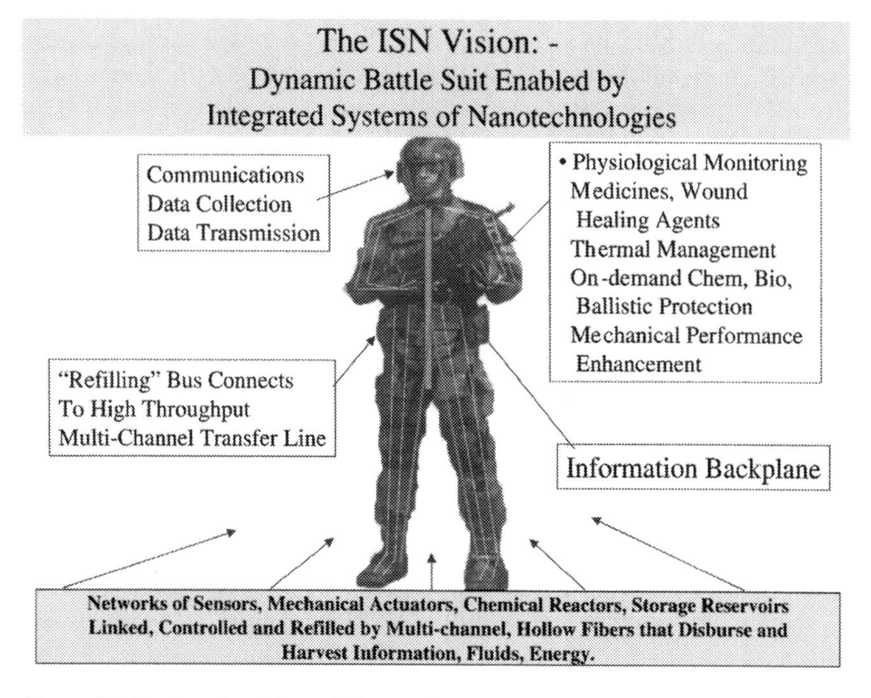

Figure 3.2 Battle-suit vision of the Institute for Soldier Nanotechnologies. (Provided by ISN, reprinted by permission.)

wavelengths that can be interrogated remotely for identification of friend or foe. Whether such a suit will allow leaps over six-metre walls and whether the total carrying load of a soldier can be reduced from above 50 to 20 kg in ten years, as suggested in first articles (Leo 2002; MIT News 2002), is open.

3.1.7 Center for Nanoscience Innovation for Defense

In December 2002, the Center for Nanoscience Innovation for Defense (CNID) was founded at the University of California (UC) (UCR 2002). $13.5 million has been given to the UC institutions at Santa Barbara, Los Angeles and Riverside; a second instalment is expected resulting in more than $20 million over three years. Additional participants include national laboratories, in particular Los Alamos National Laboratory, and ten industrial partners (Boeing, DuPont, Hewlett Packard, Hughes Research, Motorola, NanoSys, Northrop Grumman, Rockwell Scientific, Raytheon and TRW). Co-ordinating with recently founded Californian NT centers, the CNID funds will be used for state-of-the-art instrumentation and graduate fellowships. 'With disappearance of basic science research in industrial laboratories', a network is to be formed to keep the companies 'informed of the latest developments in science and technology. It's all about enabling America's business-contractors for defense technology – to keep abreast of current information' (UCR 2002). The areas to be studied at UC Riverside are: nanoscale electronic devices, spintronic devices, multiporphyrin molecular memories, neurones and nanotubes, and sensors.

3.1.8 Activities for introduction into the services

In order to accelerate utilization of near-term improvements from NT R&D, the Tank-automotive and Armaments Command – Army Research, Development and Engineering Center (TACOM-ARDEC) of the US Army at Picatinny Arsenal NJ has founded a Manufacturing, Research, Development, and Education Center for Nanotechnologies. TACOM-ARDEC is responsible for explosives, warheads, munitions, weapons, fire control and logistics. Focal areas of the NT centre are:

- advanced electronics for smart munitions,
- high-performance light-weight structural materials and processes (warhead components, gun system components, penetrators and armours),
- reactive materials and smart compounds (more powerful energetics, gradient coatings, non-conventional target effects),
- fuse components.

Guided by a 'nano-valley' vision, a high-tech industrial park is to develop military and dual-use technology in public–private partnership (Devine 2002).

3.2 Other countries

In the course of this project, systematic collection of data on military NT R&D could not be done – and would probably have been impossible in many countries anyway. Consequently, this section presents the information that was found in easily accessible sources.

3.2.1 Germany

In Germany, military activities in NT have not really begun. As of the beginning of 2003, the Federal Ministry of Defence does not fund any research or technology activities in NT in a narrower sense.[12] It has tasked the Fraunhofer-Institute for Technological Trend Analysis, Euskirchen (INT), to investigate the 'Utilization of Nanotechnology in Military Technology' (see also INT 2003). This study started in spring 2000 and is now to be finished in 2004, at a cost of about €0.5 million.[13] It will contribute to prepare decisions of the Defence Ministry about potential future activities in NT.

A 2003 study of the government-sponsored thinktank Stiftung Wissenschaft und Politik deplored gaps in utilization of new micro-technologies for security and defence and pointed out the difficulties for arms control, verification and export controls. It recommended improved protection of persons and infrastructure against misuse of miniaturized weapons, including microbiological agents (Geiger 2003).

It has to be noted that it is German policy in military technology to rely on results of civilian research wherever possible. Military funding goes only into those research and technology-development activities which are not sufficiently being done in the civilian realm.

3.2.2 United Kingdom

In the UK, some information on military NT activities is available. A general overview was published in 2001 by the Ministry of Defence (MoD) (UK MoD 2001a; see also Section 1.5.9). It has formed a Nanotechnology Panel comprised of MoD staff led by the Chief Scientific Advisor and four professors from UK universities who specialize in different fields of NT. NT research has been funded through the Corporate Research Programme at £1.5 million (€2.1 million) in 2001; slight increase in the short term was foreseen. Because of the large world-wide investment, a major NT research programme is not necessary. The MoD will carry out a Technology Watch to find out the benefits for UK defence as well as threats against the UK and its allies. Some NT research has to be funded to be accepted as a partner in international information sharing and collaborative research. This includes working closely with UK universities.

More detail was provided in a conference presentation (Burgess 2002). The wide-ranging NT interests of the MoD include:

- power sources; alloys, polymers, composites, textiles; explosives, pyrotechnics, propellants; self repair systems; weapons (intelligent, autonomous, accurate); stealth and counterstealth;
- secure messaging; global information networks, sensing;
- vaccines, medical treatment; wound repair, decontaminants; chemical/biological protective creams; 'lab on a chip' chemical/biological agent sensors.

Also, the potential of 'unethical use' leading to new biological and chemical weapons is mentioned.

Since '98% of R&D relevant to defence is funded from elsewhere', mixed funding from the MoD and Research Councils goes to NT Interdisciplinary Research Centres (IRCs) and to UK academia, and from the MoD to the firm QinetiQ[14] and to industry. From the MoD Corporate Research Programme, the following examples were mentioned: nanocrystalline materials; polymer nanocomposites; nanostructured carbon and hydrogen storage; nanostructured materials for sensors, random-access memory (RAM), etc.; quantum coherent devices; molecular electronics; nanophotonics.

QinetiQ has founded QinetiQ Nanomaterials which grew out of work on energetic materials using 100-nm particles. After various trials, a production facility is being commissioned. One product is lead-free ignitors. QinetiQ, claiming to have the largest nano-materials and NT group in Europe (150 people), is active in the areas of: hybrids, nano-sensors, nanomagnetics, biomimetics, nano-carbon and nanoelectronics (Reip 2002; QinetiQ 2003).

At least three UK universities have started projects on nanostructures in polymer matrices, sponsored by the US Army European Research Office (Shaw 2002; see also Section 3.1.3.2).

3.2.3 Other West European countries, European Union, NATO

In *France*, a nuclear-weapon state with a strong tradition of military R&D, there are indications that significant efforts in NT have begun. The second International Meeting on Micro and Nanotechnologies Minatec 2001 in Grenoble was supported, among others, by the French Defence Procurement Agency (Délégation Générale pour l'Armement, DGA); one 'regional day' organized by DGA was devoted to 'Science and Defence', with contributions from the Commissariat à l'Energie Atomique (CEA) which is also responsible for nuclear weapons (however, the successor conference 2003 had no such sessions) (Minatec 2001, 2003). Unfortunately, efforts at getting information about the budget or topics of French military NT R&D were unsuccessful.

On the *Netherlands*, no specific information was available. Judging from the activities of the Dutch Organization for Applied Research – Physics-Electronics Laboratory (TNO-FEL) in the area of MST (Altmann 2001: Section 3.3), one can assume that work in NT will be increasing.

In *Sweden*, an NT programme is being launched that is motivated on the one hand by the need for advanced equipment, but on the other hand by the needs for retaining an advanced defence industry base, an attractive R&D base and a strategic level of competence. The Swedish Defence Research Organization FOI has started a planning process for projects which should integrate industry, universities and defence. A first workshop has taken place in March 2002, a programme decision was due in March 2003, the first and second phases should start in July 2003 and July 2005, respectively. Technology demonstrations are foreseen for autumn 2008 (Savage 2002; on FOI, see http://www.foi.se).

In the *European Union*, the Framework Programmes for Research, Technological Development and Demonstration Activities up to now do not include military R&D (EC 2004). However, in the preparations for the Common Foreign and Security Policy, security and defence research is being increasingly included. While NT is not (yet) explicitly mentioned, it will probably play a role in efforts directed at 'technology solutions for threat detection, identification, protection and neutralization as well as containment and disposal of threatening substances including biological, chemical and nuclear ones and weapons of mass destruction' (EC 2004). In the European Defence Industry Group (EDIG) a Technical Case Study on NT applications has been proposed which could lead to EUCLID programmes (Burgess 2002).

In the *North-Atlantic Treaty Organization* (NATO), there has been a Task Group 'MEMS Applications for Land, Sea and Air Vehicles' from 2000 to 2003, dealing with MST.[15] A similar group for NT may be formed in 2004.

3.2.4 Russia

With the limited amount of effort that could be afforded in the present project, no reliable information on military R&D of NT in Russia was found. Overview articles on NT work at large make clear that there is a wide range of civilian activities, even though hampered by economic difficulties (e.g. Holdridge 1999; Andrievski 2003 and refs). Russian institutions take part in many international collaborations; there is, for example, a co-operation agreement between Russia and the EU for participation in the 6th Framework Programme of the EU.[16]

In its National Security Concept of 2000, the Russian Federation stresses its concern, on a general level, with 'the growing technological surge of some leading powers and their growing possibilities to create new-generation weapons and military hardware'. Among the principles for use of military force, should that become necessary, it is stated that 'the restructuring and conversion of the defence industries should not come into conflict with the creation of new technologies and research-technical possibilities, the modernization of weapons, military and specialized hardware, and the strengthening of positions of Russian producers on the world market of weapons' (Russia 2000; see also Russia 2000a).

With a long tradition in military high technology and active NT R&D, there can be no doubt that Russia will be capable of using NT in various ways in the armed forces, should this become a high priority.

3.2.5 China

No hard information on military R&D for NT in China was found. NT research in general is very advanced in China; centres have been set up in the Chinese Academy of Sciences and various universities. Chinese institutions participate in international collaborations, and international conferences have taken place in China. The government has set up a National Coordination Committee for nanoscience and nanotechnology; in the list of participating ministries and agencies, no defence-related institution was given explicitly (Bai 2001). However, it is probably safe to a assume that the wording 'and so on' comprises the Ministry of Defence or the Commission of Science and Technology for National Defence (COSTIND) (see COSTIND 2002).

Hints at alleged Chinese plans to attack the USA by 'microscale and ant robots' that were contained in a US overview of Chinese military literature have turned out as a misrepresented report about a corresponding US study (see Section 3.4).

Relatively active in basic NT research and with expanding activities in military high technology at large, China is certainly able to develop all kinds of military applications.

3.2.6 Other countries

Of the more than thirty countries with NT activities or plans, or the fifteen most active in publications or patents (see Section 2.5), many will nearly exclusively focus on civilian products and markets. For example, *Taiwan* and *South Korea* will likely continue their traditional economic path into the NT era (e.g. Lee C.K. *et al.* 2002; Lee J.-W. 2002).

Japan is one of the biggest players in civilian NT, but has not been that active in military high technology in general. This will probably hold for NT in the future unless general Japanese policy were to change.

In *Israel*, on the other hand, calls for founding a largely commercial NT initiative are having military connotations from the outset: former Prime Minister S. Peres mentioned the possibility of military units without soldiers and noted the importance of Israel's nuclear option. The NT Committee established by the President of the Israel Academy of Sciences that called for a five-year Israel NT Programme had one member from the Ministry of Defence and mentioned military development (Peres 2003; Netfirms 2001).

In *Australia*, the Defence Science and Technology Organization (DSTO) has prepared a first overview study on potential military applications of NT, with a view on the future land force; this effort will continue (Wang and Dortmans 2004).

For *India*, active in military high technology, possessor of ballistic missiles and nuclear weapons, one can assume that military R&D will soon turn towards NT.

The other South Asian nuclear state, *Pakistan*, might follow.

Of the list of states of concern to the USA (Cuba, Iran, Libya, North Korea) none is remarkably active in NT at all, so indigenous development of military NT systems can practically be excluded for the foreseeable future. Of course, biomolecular research is possible for all of them already today, and growing availability of NT tools and methods will provide increasing capabilities for many state and non-state actors in the future.

3.3 International comparison of military NT efforts

Detailed figures on military R&D efforts for NT exist only for the USA. For 2003, its expenses (within the NNI) were $243 million (see Table 3.1). Some information was given above for the UK and Germany. For comparison, one can make a cautious guess about 2003 expenses. Assuming that the UK funding of about €2.1 million in 2001 were scaled up by a factor of 1.5, one arrives at €3.2 million. For Germany, it seems reasonable to assume a continued expense at the level of the previous years, around €0.2 million.

In the absence of more information, one can assume that NT-related military spending in France and the Netherlands is similar to that in the UK. In other European countries, it will be less. The sum across Western Europe is probably below €15 million and almost certainly under €20 million per year. This would mean that the US spending is above twelve to sixteen times the West European one.

Speculating about Russia's and China's expenditures, one can note that the overall military R&D budgets of the two other official nuclear-weapons states UK and France are about $4 billion and $3 billion per year, respectively, so that the NT-related expense given or estimated above is about one tenth of a per cent (it is 0.6 per cent for the USA with about $40 billion total). Assuming a similar ratio for Russia (total military R&D about $2 billion) and China (roughly $1 billion), one arrives at NT-related figures of $2 million and $1 million, respectively (budget figures for 1999 (see BICC 2001), the China estimate is from 1994 (Arnett 1999)). Doubling or tripling would raise the numbers to the level of the UK. Summing all the mentioned countries and allowing 10–20 additional ones with on average $2 million/year would yield a global expenditure outside of the USA of between $30 and $40 million per year. If that were true, the ratio between the USA and the rest of the world would be between 8:1 and 6:1.

As a more cautious estimate one can assume that the present spending ratio is between 4:1 and 10:1. However, the very small relative portion in the military R&D budget in all countries indicates that there is considerable leeway for increases – and experience suggests that expenses will strongly increase as technologies move from research to development, and again from there to acquisition and deployment.

3.4 Perceptions driving an NT arms race?

From the Cold War, we know that threat perceptions are one of the important drivers in the quest for new military technologies. Bomber and missile gaps seen in the competition with the Soviet Union were important arguments against restraint in the USA; only after U-2 spy planes and observation satellites had arrived was it found out that these gaps did not exist. Even though the fundamental ideological confrontation is over, the basic mechanisms are still at work in the relation between potential military opponents – especially for the USA, these include Russia and China. Military secrecy on their part together with worst-case assumptions on the US part could well cause mutually accelerating threat perceptions that drive an NT arms race.

One example how this could come about is provided by an overview on Chinese military thinking provided by M. Pillsbury, a US China specialist and former DoD official. In his detailed account how 'China Debates the Future Security Environment', there is a section on the Revolution of Military Affairs. Here he writes:

> *Nanotechnology Weapons*
> An article by Major General Sun Bailin of the Academy of Military Science is particularly important because it illustrates how asymmetric attacks on US military forces could be carried out with extremely advanced technology. General Sun points out that US dependence on 'information superhighways' will make it vulnerable to attack by microscale robot 'electrical incapacitation systems.' (635)
>
> The targets would be American electrical power systems, civilian aviation systems, transportation networks, seaports and shipping, highways, television broadcast stations, telecommunications systems, computer centers, factories and enterprises, and so forth. Sun also suggests that US military equipment will also be vulnerable to asymmetrical attack by 'ant robots'.
>
> According to General Sun, these are a type of microscale electromechanical system that can be controlled with sound. The energy source of ant robots is a microscale microphone that can transform sound into energy. People can use them to creep into the enemy's vital equipment and lurk there for as long as several decades. In peacetime, they do not cause any problem. In the event of relations between two countries deteriorating, to the point that they develop into warfare, remote control equipment can be used to activate the hidden ant robots, so that they can destroy or 'devour' the enemy's equipment.
>
> (Pillsbury 2000: Ch. 6)

Ref. 635 is to the English translation, printed in a book also edited by Pillsbury, of the original article by Sun Bailin (published 1996 in National Defense (China)) (Sun 1996/1998). Sun does not mention attacks on the USA at all – his article was a report to a Chinese audience about a 1993 study from the US RAND Corporation on military applications of microelectromechanical systems (Brendley and Steeb 1993). In the RAND report, microscale robots and insect platforms were mentioned as potential future weapons of the USA. Potential enemy and terrorist use against the USA was only mentioned in a short paragraph.[17]

This obvious misrepresentation points to a potential negative feedback cycle. The high importance that the USA attaches to military NT and the general transparency about its R&D efforts must make strong impressions on Chinese military planners, so that they will have high motives to increase their military-NT activities even if only for 'defensive purposes'. Together with the relatively strict secrecy in China, sporadic observations and reports tend to become exaggerated in the USA, increasing the perceived need to advance even faster.

The same mechanism is certainly at work with Russia, and may in the future hold for several other NT-capable countries, too. Thus, without international constraints the future may bring strongly increased military NT-R&D activities and the deployment of NT-based military systems in many countries.

There is a way out of this problem: it requires increasing transparency on military R&D on the part of Russia, China etc., and giving up the aim of absolute military-technological superiority on the part of the USA. Concepts how international limitation of the most dangerous military NT applications could be carried out will be presented in Chapters 6 and 7.

4

POTENTIAL MILITARY APPLICATIONS OF NANOTECHNOLOGY

The main goal of this chapter is to describe various potential military nanotechnology (NT) applications and to estimate the time frames for their potential introduction. This is done for those areas that can be extrapolated from present science and technology in Section 4.1; a summary is given in Section 4.2. Potential military applications of molecular NT are considered in Section 4.3. This discussion is much more speculative, of course, but it is needed because MNT cannot be excluded on the grounds that it violates the laws of nature. The final Section 4.4 casts a short look on potential countermeasures against military NT systems.

4.1 Military applications of NT

The presentation of potential military applications starts with the more generic ones (e.g. computers), then proceeds to those that are rather specific to the military (e.g. weapons). In compiling the applications, I have on the one hand used existing sources of various types; systematic overview presentations do not yet exist. On the other hand, I have contributed my own analysis. Not everything discussed need be under military R&D at present. On the other hand, the list is certainly incomplete. Additional applications may have been discussed in secret, others may only come into view as NT evolves further. The presentation is intended as an overview; in most cases, some relevant references are given that can be used to dig further. Where possible, I mention examples from US R&D, in particular by DARPA.[1] For several reasons, however, not much is publicly available about NT work for new weapons of mass destruction.

The times to potential introduction have only been estimated roughly; they depend on many factors anyway, not the least the amount of funding, but one also has to expect surprises – on the one hand, unexpected breakthroughs or cross-fertilization from a different area or, on the other hand, unanticipated obstacles. Thus, the times are only given in coarse categories:

- within the next five years,
- five to ten years from now,
- ten to twenty years, and
- more than twenty years from now.
- In addition, there is a 'speculative' category for applications which cannot be excluded on the grounds that they violate the laws of nature.

4.1.1 Electronics, photonics, magnetics

In *microelectronics*, NT will allow continued reduction of component size, later using new principles such as nanotubes or bio-molecules. Quantum effects will dominate below some scale, but could be used intentionally. Devices utilizing the electron spin could be fast at low-power consumption. Continuing from the micro scale, NT can provide not only smaller components such as transistors, but also mechanical resonators for filters at gigahertz frequencies.

In *photonics* – generation, transfer, switching and processing of optical signals – NT provides possibilities for sources, detectors and devices such as waveguides, filters, couplers, modulators, etc. The optical frequencies of absorption and emission of quantum dots can be tuned (see Section 2.1.4). Photonic crystals are structures with periodic variations in index of refraction where – similar to electrons in a crystal – light cannot propagate at certain frequencies and in certain directions; this can be used for waveguides, switches etc. (AFOSR projects: Pomrenke 2002: 44–50; NRL work: Kafafi 2003: 12–17). Integrated microelectronic-nanophotonic circuits can lead to fast data rates (Terabit/s = 10^{12} bit/s) at small size and low power (Pomrenke 2002: 44).

For *displays*, NT provides several possibilities. One is using field emission to release electrons in a vacuum tube – heating of the cathode is no longer needed. The electric field is particularly strong at sharp tips, e.g. those of carbon nanotubes. An array of nanotube emitters facing individual phosphor pixels would also obviate the need for scanning the electron beam across a screen; the device could be flat. Such displays would still need vacuum, but would use much less power and could be much smaller than present vacuum-tube displays (NRL work: Colton 2003: 26). A second concept uses organic molecules that emit bright light of selectable colour. Such displays could have pixel sizes of 12 μm and below, would be thin, very lightweight and flexible, would last a long time and operate at a very wide temperature range (NRL work: Kafafi 2003: 5–11). They could be helmet-mounted or integrated into battlesuits.

Nanomagnetics (Goronkin *et al.* 1999; Koch 1999) made a great leap with the discovery in 1988 of giant magnetoresistance (GMR) – the decrease of electrical resistance between thin ferromagnetic and nonmagnetic layers when a magnetic field is applied. In 1997, GMR sensors were introduced for hard-disk read heads, allowing higher storage densities. With one very thin (<2 nm) insulator layer between two ferromagnetic layers, the tunnel current depends on the mutual magnetic directions; such magnetic tunnel junctions can be used as memory cells. Different from dynamic random-access memory (RAM), magnetic RAM would not require constant refreshing, and would keep its contents even when the control circuit is switched off, allowing immediate start of a computer without a tedious boot process (DARPA Spintronics 2003). Bulk material made from nanocomposite magnets will allow stronger permanent magnets that would increase the energy extracted from magnetic motors or generators (DARPA MetaMaterials 2003).

Progress in microelectronics, photonics and magnetics will combine to enable much more capable computers and communication links. Even further miniaturization at reduced power consumption and faster processing rates is promised by qualitatively different technologies. *Carbon nanotubes* could be used as conducting wires, two cross-linked tubes can form a transistor (Lieber 2001). In *molecular electronics*, molecules are used to conduct electric current, as switches or as memory cells. In the corresponding DARPA program, scalability of logic devices to 10^{12} gates/cm^2 must be demonstrated; memory must achieve a volume density of 10^{15} bits/cm^3 (DARPA Mole 2003; see also Chen *et al.* 2003, but also Service 2003). The former is four orders of magnitude above the present state in production of two-dimensional CMOS structures.

Making use of NT-enabled miniaturization and integration, complete electronic systems could fit into a cubic millimetre or less. Power supply will present problems, however – batteries will not shrink in parallel. Where only the microcircuit proper needs to be fed, a millimetre-size power supply could suffice, possibly using non-traditional concepts (see Section 4.1.5). In many applications, however, the system will need to communicate by radio to some distance. This will need power supplies as well as antennae of larger size,[2] e.g. centimetres – these components can limit the scaling down of systems. Sizes of cm and masses of tens of grams are unproblematic for systems worn by persons, or in equipment of macroscopic size. Sophisticated electronic systems are to be expected everywhere throughout the military, not only in transport containers, pallets and boxes, but also in very small pieces of equipment, including rifles, garments, glasses and ammunition.

Because of the variety of applications, estimates for introduction times are difficult to make. Many categories will advance continuously from the present through the next twenty years and beyond. First flexible displays and spin-effect electronics could arrive in 5–10 years. For nanotube-based and molecular electronics the expected time is rather 10–20 years.

4.1.2 Computers, communication

Computer components will continue to become faster and more capable. For more than a decade, advances will be based on evolutionary adaptation of microelectronic processes to smaller structure sizes. Continuation of Moore's law beyond 2015–2020 will require new principles, e.g. based on nanowires or molecules. Hard disks will be replaced by different forms of nanoscale memory without rotating parts. New computing paradigms – solving complex problems by massive parallelism of DNA molecules, quantum computers or maybe neural architectures – may also contribute (DARPA Biocomp 2003; DARPA QuIST 2003).

NT-based electronics and photonics will allow higher transmission frequencies and communication bandwidths; optical transmission to some distance requires either a fibre or a free line of sight.

While computing power will grow over about four orders of magnitude in the next twenty years, the sizes and power requirements will shrink drastically. The price per computing power will fall, too; depending on the production processes, even the absolute price of one processor may decrease markedly.

With the performance of a present-day PC packed into a volume that counts by cm^3 or even mm^3, such processors will be used throughout the military, embedded in practically all durable components, from vehicles via rifles to uniforms, and in most throw-away items, such as munitions. As required, they will be integrated with communication units, displays, input units (touchpads or microphones for speech), small sensors and actuators. This miniaturization makes possible qualitatively new applications such as micro-sensor systems, guidance systems even in small projectiles, implanted processing and communication devices or small robots (see Sections 4.1.10, 4.1.14 and 4.1.16, respectively). These ubiquitous computing and communication devices will be integrated in and form networks that are flexible and reconfigure themselves continuously. This holds not only on the battlefield, but also in logistics, where single items such as boxes are likely to be equipped with systems that monitor and record what happens (acceleration, temperature, humidity etc.) and communicate autonomously with the control systems of the respective warehouses, transport planes etc.

On the other hand, where there is no need for massively shrinking size, computers will become all the more powerful. Modelling of phenomena in weapons, materials and organisms will advance markedly. In training, more complex situations can be covered and represented by virtual reality. Given appropriate progress in software and artificial intelligence (see Section 4.1.3), automated planning, decision preparation and management will comprise combat on all layers, from the platoon to the highest strategic level. Probably, there will be a sliding transformation to increasingly autonomous decision-making. Similar developments are likely in logistics and planning in peacetime.

If NT-enabled quantum computing becomes practical, it would be applied on the one hand for secure communication, on the other for faster breaking of codes. In addition, it would be used in many fields to solve hard optimization problems or to run simulations of matter phenomena.

Computer performance will probably continue to improve, based on various forms of NT, more or less according to Moore's law for at least twenty years. New military applications will arrive throughout this period. When quantum computers could become practical, is difficult to predict – it may well take more than twenty years.

4.1.3 Software/artificial intelligence

Software – concepts, algorithms, architectures etc. – is to some extent independent of hardware. On the other hand, vastly increased processing power allows new concepts to be implemented, and stimulates software development, as evidenced by the addition of sound and images to formerly text-based PCs. The same will apply for NT-enabled hardware improvements.

One important question is whether or when computers will be able to reach human-like levels of reasoning. In the field of artificial intelligence, early expectations of fast breakthroughs in pattern recognition, chess-playing, natural-language processing etc. – feats that seem easy for a human – were disproved over and over. However, even though it took much longer than expected, impressive successes have been achieved today.

As shown in Section 3.1.2.3, present DARPA programs aim at human-like interaction and reasoning. Even if one is skeptical about the farther goals, one does not need much imagination to expect computer systems that

- communicate regularly in spoken natural language,
- translate in real time between different languages,
- provide vast databases and search engines that make use of everyday knowledge,
- visually recognize their environment and are capable of navigating in it as well as manipulating it.

Large-scale systems will exist for strategy planning, battle management and logistics. The operation of armed forces will be characterized by many autonomous decisions on all levels. Uninhabited vehicles and robots of macro- and microscopic sizes could become routine. The systems would adapt and learn from their experiences as well as from the exchanges with human operators.

Software progress will continue and lead to new levels of artificial intelligence – so that the latter finally will deserve its name. Whether the so-called software bottleneck will continue to hamper development is difficult to predict. For the time being, software and computers that could form their own goals and plans, and then go about implementing them, are only a possibility in principle. It is much easier to predict that the raw processing power of a PC will reach and transcend that of the human brain (taken as around 10^{16} operations/second)[3] by about 2020 (e.g. Williams and Kuekes 2001; Kurzweil 1999: Ch. 6). Whether this achievement will quasi-automatically result in human-level intelligence, or whether a significant breakthrough in software technologies (or understanding of brain function) will be needed, is open. Computer experts active in the US NNI, at least, have predicted that by 2020 a limited Turing test (five-minute conversation) would be passed (Williams and Kuekes 2001); Williams (2002) accepted the Turing prediction that machines would finally become sentient.

Step-by-step software improvement will go on all the time, parallel to hardware development. Whereas first forms of general natural-language communication and autonomous complex behaviour of robots may arrive in 5–10 years, fully developed capabilities will probably need 10–20 years. 'Seamless integration of autonomous physical devices, computation software agents, and humans' is not likely to occur before twenty years' time. Whether human-like intelligence could be achieved significantly later than twenty years from now, and if so, when, one can only speculate. Past experience suggests humility concerning such an achievement, but surprises and leaps cannot be excluded.

4.1.4 Materials

NT provides a wide range of materials with improved as well as new properties. This concerns structural materials (mainly composites and metals), functional and active materials.

Composite materials have been used in military systems for a long time, e.g. glass- and later carbon-fibre reinforced plastics in aircraft for high strength at low weight. *Composites with nanoscale additives* promise significant improvement in many respects (Rudd and Shaw 2001; Busbee 2002: 27–37). For example, clay-derived layered silicates in thermoplastics have led to reduced flammability; application could be in power-cable insulation. A doubling of the elastic modulus and hardness has been observed, tensile strength and fracture toughness can increase by 50 per cent; water and solvent permeability can decrease by a factor of 10. Hybrid hard/ductile layered polymers can be used for improved lightweight transparent armour. Using carbon nanotubes as additives can turn a composite material into an electrical conductor. It is unclear at present if nanoparticle composites can go beyond long-fibre composites, but adding nanofibres may improve the latter. In aerospace, conductive plastics could be used for signal wires, against static discharge, for actuators or for flexible circuits. Multifunctional plastics could be applied in propulsion, deployable space structures, tanks or tyres.

Carbon-nanotube composites that would make use of their extreme tensile strength would provide a qualitative jump in material strength at much reduced weight (see Section 2.1.5). Should a 'space tower/elevator' extending beyond the geostationary orbit (see Section 2.1.5) ever be built, military operations would profit from it by easy and cheap access to space.

Metal with nanoscale crystallites can show markedly improved properties (Koch 1999; Busbee 2002: 38–39). Whereas with cooling from a molten phase, grain sizes of several μm and above result, consolidation of nanosized powder can keep the crystallites much smaller (special efforts are needed to avoid pores). While the elastic moduli remain about the same, the strength and hardness have been found to grow as grain size decreases. Hardness increased typically by a factor of 2–7. With a nanophase aluminium alloy, a 60 per cent strength increase was possible. Smaller grains can make materials more brittle and tougher. Greater hardness and toughness can be used in *nanostructured coatings* which exhibit low friction and wear at high and low temperatures (Busbee 2002: 40–42).

Amorphous metal that has an irregular structure on the nanoscale, but is not usually produced using NT, can support about double the elastic strain of traditional multicrystalline metal (Christodolou 2000; DARPA SAM 2003; DARPA Fact 2003; for use in penetrators, see Section 4.1.12). Tensile strength and hardness are about twice the values of steel, fracture toughness is three times as high. Because toughness increases with strain rate (speed of deformation), improved resistance against explosive blast and ballistic impact may result. Reduced wear and corrosion are possible, too. For naval applications, iron-based material is being investigated, for airframes and space structures the Air Force Research Laboratory (AFRL) is evaluating aluminium- and titanium-based amorphous metal.

Concerning structures of very small objects, biology has provided concepts for materials that are composed of simple building blocks and self-assemble in a hierarchical way to three-dimensional structures on the nano- and micro scale. For technological application, one can use DNA molecules with branches; by producing appropriate fitting patterns of bases, they can be caused to connect in a specific way to form complicated structures in two or three dimensions (Jelinski 1999). These could be used, for example, for molecular electronics (DARPA Budget 2002: 3–11).

In the category of *functional materials*, one example is provided by nanocomposite permanent magnets that could provide higher efficiency for power electronics and electric motors (DARPA MetaMaterials 2003). A different type is a self-healing capability provided by nanoscale additives within certain materials (Busbee 2002: 32). Nanofibres could be used for textiles with integrated sensors and electronic connections; membranes with nanoscale pores could be used for filtering and protection.

Active materials, that is materials that can change shape or exert a force when exposed to appropriate control conditions, exist already, using, for example, the piezoelectric or the shape-memory effect. They can move control flaps on munitions or the legs of robots. NT can improve the properties as well as lead to new kinds of active materials such as contracting molecules moving an exoskeleton (see Section 4.1.6). Alternatively, biological or hybrid materials could be used, based, for example, on myosin or kinesin. Integrated with sensors, power and processing, active materials can become *'smart' materials*, adjusting stiffness to reduce vibration or changing form for different aerodynamic characteristics (DARPA Smart 2003).

Not all material enhancements hoped for need to work out or become economically feasible. Nevertheless, changes of revolutionary character are to be expected. NT-enabled improved materials will likely be used throughout the military. Time frames for success are difficult to assess. Routine use of nanophase composites and metals could begin in 5–10 years whereas introduction of high-strength carbon-nanotube-reinforced plastics could be 10–20 years off. NT-based smart materials may take 5–10 years to first actual use; biologically inspired materials may arrive after 10–20 years, unless a significant breakthrough occurs.

4.1.5 Energy sources, energy storage

In the area of energy conversion and storage, NT will on the one hand provide macroscopic systems with higher efficiency at lower mass, on the other hand make available new microscopic devices for applications where very low power is sufficient.

In the first class, fuel cells with nano-structured electrodes and membranes are probably most important. Medium-sized cells could be used for uninhabited air vehicles and in outer space. Larger types (tens of kW and above) could be used for land vehicles and ships. Should NT provide a means for efficient hydrogen storage, this, together with a fuel cell, would be an important contribution to an all-electric vehicle.

Organic nanocomposites could provide flexible, very lightweight solar cells (DARPA Budget 2003: 299); they could be applied wherever power demand is small, or a large area is available.

NT concepts can also be applied to small power generators which are based on microsystems technology (MST), such as those being developed in the DARPA Micro Power Generation and Small-Scale Propulsion Systems programs (DARPA MPG 2003; DARPA SPSS 2003). Using hydrocarbon fuels, these micro thermo-electric converters, combustion engines, fuel cells and fuel reformers are to provide tens of μW to about a watt at more than ten times the energy density of the best currently existing batteries. Larger fuel cells of 60 or 20 W could power soldier systems or small robots (prototypes are being developed by DARPA (DARPA Energy 2003)). NT could improve these generators, for example, by stronger, lighter and more heat-resistant materials, by wear-reducing layers or by nanoporous membranes.

If a device is worn by a person, power may be gained from motion, small thermal gradients, or reactions of biochemical molecules. In one DARPA/ONR project, an implantable bio-fuel cell is being developed that would be mounted in a blood vessel; by glucose oxidization and oxygen reduction at a special anode and cathode, respectively, more than 1 μW power was produced over a week in a laboratory test. Such a fuel cell could be used to power micro-sensors, -actuators and telemetry devices in plants, animals, or humans (DARPA Energy 2002; Mano *et al.* 2002).

While a few NT-based energy technologies (maybe some solar cells) could appear within five years, most are 5–10 years off, some 10–20 years.

4.1.6 Propulsion

In traditional engines of the reciprocating or turbine type, NT-based materials or surfaces may allow higher temperatures to be used which will lead to higher thermodynamic efficiencies. Further improvement is likely to result from lighter weight of moving components and reduced friction from hard surface layers. This can be accompanied by reduced wear. Together with MST, very small versions of such engines may become practical, as mentioned in Section 4.1.5. Such miniature engines could power miniature (land, water, air) vehicles and robots.

At some small scale, electromagnetic and in particular electrostatic motors may become more practical. Also here, NT could in several respects improve MST motors. NT and MST may also allow macroscopic efficient electric motors that could be used to drive future cars etc. Situated in the wheels and used for braking, too, they would obviate the need for gearbox, clutch, dynamo, etc., in an all-electric vehicle.[4]

Different principles could be used in biologically inspired systems. Shape-changing materials, contracting molecules etc. can be used to move legs or flapping wings, on size scales from micro- to macroscopic. At the Institute for Soldier Nanotechnologies (see 3.1.6), the polymer polypyrrole, which contracts or expands depending on electric state, is being investigated for use in an exoskeleton (ISN 2 2002; Talbot 2002). To propel a bacterium-size system, the natural nanomotor could be used to rotate a flagellum. Alternative molecular motors could be the biological F_1-ATPase molecule (present in mitochondria membranes) which is driven by adenosine triphosphate (ATP), or, for linear motion, myosin and kinesin (Jelinski 1999; a hybrid artificial/biological ATPase rotator was created – with funding by ONR, DARPA and civilian agencies – by Soong *et al.* (2000)). Proceeding from these natural systems, improved artificial versions could be created.

Concerning space rockets and all kinds of missiles, NT-improved materials, housings, structures, tanks, pumps etc. could be made markedly lighter. Char-forming nanocomposites could be used as ablative insulation in solid-rocket motors (Busbee 2002: 34). Very small engines could drive similarly small rockets; one DARPA program, not yet using NT, wants to demonstrate a liquid-fuelled micro-rocket with turbopumps, providing 15 N thrust (the weight force of 1.5 kg mass); such rockets could deliver 200 g satellites to low earth orbit (DARPA SPSS 2003). (For rocket propellant see Section 4.1.8.)

NT-enabled improvements in traditional engines will probably arrive in 5–10 years. Miniature and microscopic engines may take a similar time, or become practical only after ten years. The same holds for NT-based electric motors and for NT-enhanced rocket engines (large and small). Biologically inspired motors are likely to be applied routinely in 10–20 years.

4.1.7 Vehicles

Materials that are lighter and stronger, engines that are lighter and more powerful and more efficient energy storage will allow construction of military vehicles of traditional types with markedly improved characteristics. This holds on the ground, in and under water and in particular for aircraft and space vehicles, where weight reduction is of highest importance (for outer space, see Sections 4.1.6 and 4.1.18).

Among ground and water vehicles, the mass reduction would be most pronounced for those used for transport purposes. NT could bring new options of providing them with light armour. Since NT will probably not be able to substitute heavy armour (decimetres of steel), tanks and battle ships will change less. (See Section 4.1.11.) With efficient fuel cells and better storage provided by NT, propulsion could change to electric.

For aircraft, mass could be reduced further if shape-changing materials obviated the need for control flaps and the associated hydraulics. Reduced mass would allow less powerful engines or higher agility – however, already at present it is the pilot's tolerance that limits the operational acceleration – higher payload, and/or longer range.

More drastic mass reduction would ensue with uninhabited vehicles, for which NT-based computers could provide the required degree of autonomy. Again the effect would be strongest for aircraft. Reduction would find its limits in the payload, however, as long as bombs of 1,000 kg and hundreds of traditional machine-cannon rounds will need to be carried. This could change only if much smaller, precise weapons/ammunitions of sufficient destructive effect were to become possible. The existing DARPA programs for uninhabited combat and non-combat vehicles have not yet embraced NT (DARPA TTO (2003) lists nine programs for ground and air systems). There is no doubt that as soon as NT R&D deliver applicable results, these will be included fast.

Uninhabited autonomous vehicles can be seen as first stages of robots of macroscopic size. MST and NT will make possible very small vehicles and robots, see Section 4.1.16.

Whereas NT-based upgrades in control electronics will begin to be used already within five years, vehicles making some use of NT in structures, flow control, engines or energy storage will probably enter service within 5–10 years; systematic use is expected after ten years.

4.1.8 Propellants and explosives

Energetic materials contain a fuel and an oxidizer that on ignition react with each other, releasing the chemical energy and creating hot gases at high pressure. These can perform mechanical work – flow fast out of a nozzle and accelerate a missile by recoil, propel a projectile in a gun barrel, accelerate fragments or produce a blast wave. Fuel and oxidizer can be separate compounds which are mixed for fast reaction, or they can be combined in a single molecule such as trinitrotoluene (TNT, H_2C-C_6H_3-$(NO_2)_3$). Here, nitro (NO_2) groups provide the oxygen for carbon and hydrogen; the nitrogen atoms form N_2 in the detonation. In a sense, the molecule serves to keep the reactants at a distance until the intended reaction time. The energy density (also called heat of explosion, at constant pressure) of TNT, for example, is around 4.5 MJ/kg (Ullmanns 1982: Table 5).[5]

Nanoparticles allow much better mixing of fuel and oxidizer, promising much faster reaction of composite material. The associated risk of unwanted ignition during mixing can be reduced by sol-gel methods (Parker 2000a): nanoparticles are created in solution; after evaporation of the solvent, the particles ('sols') form a three-dimensional skeleton with pores ('gel'). Thermites such as iron oxide mixed with aluminium, used in airbag igniters or for welding, have been produced in this way. A monomolecular material can be grown as nanocrystals from solution in the gel. Explosives RDX and PETN were embedded in this way in a silica matrix. In a third concept, the sol-gel matrix would itself be energetic, consisting of a nanostructure where fuel and oxidizer are distributed in exact stoichiometric relation. By varying the composition, the energy density can be programmed (CMS 2002: 41).

A more radical idea would be to form new molecules of higher energy densities which are not easily accessible by traditional chemistry, perhaps even using techniques of molecular NT such as molecular assemblers. It is conceivable, for example, that nitrogen, carbon, or hydrogen atoms could be packed more densely with oxygen atoms than in TNT (of 4.5 MJ/kg) or nitroglycol $(ONO_2)_2(CH_2)_2)$ of 7.4 MJ/kg (Ullmanns 1982: Table 5). One research example is nitrogen fullerenes (e.g. $C_{48}N_{12}$) that are being modelled by computer at the US Livermore Laboratory. Assuming an oxidizer (e.g. a perchlorate or metal oxide), one could calculate an energy density. Whether new explosives gained by molecular design would be stable enough to be stored and transported is unclear, of course. Intuitively, an energy-density increase by several times 10 per cent seems possible; a doubling of energy density is probably unrealistic, however.

Similar considerations will apply to solid-rocket propellants: considerably higher – but less than a factor 2 – energy density is plausible, with better control of the burning process.

With improved capabilities to tailor explosives by NT, nanocomposites could also be used as igniter charges (Burgess 2002: 10).

NT-improved explosives and propellants can likely enter military use within five or 5–10 years.

4.1.9 Camouflage

NT provides several possibilities in principle to change the (apparent) colour of a surface. Similar to concepts for thin displays, mobile pigment particles could change position or orientation to expose differently coloured sides. Alternatively, surface structures of sub-wavelength periodicity could create colour impressions as with the scales on butterfly wings; however here the perceived colour would depend on direction. Such active camouflage could be used on battle suits, land vehicles, aircraft and ships (e.g. ARO Materials 2001). By changing the colour(s) and texture to that of the background, the person/object would merge with it. However, this applies generally only for one viewing direction – seen from another, the background normally differs, and simultaneous 'invisibility' from all directions cannot be achieved (ARO Materials (2001) wanted to 'Make soldier invisible across the EM spectrum').

An alternative is to use light-absorbing material. This is similar to radar-absorbing layers on stealth aircraft. In the visible and infrared region, a multi-layered photonic material is now being transformed into threads for a battle uniform. The reflectivity for certain wavelengths is to be tuned in real time; beside absorption, for example, for visible light, reflective patterns in the infrared could be produced which could serve as an 'optical bar code' identifying the soldier as friendly to one's own forces when using special goggles (Talbot 2002). Nanophotonic materials can also be tailored to have particular absorption properties for radar or infrared radiation.

NT-enabled active camouflage will probably become available in 5–10 years.

4.1.10 Distributed sensors

NT allows the manufacture of extremely small sensing elements for all kinds of quantities. The sensitivity can be very high: in principle, single molecules of a chemical agent or one microorganism of interest can be sensed, e.g. by a micro-cantilever with appropriate coating. A different concept would use biological sensors that provide high sensitivity and selectivity. Nanoscale interfaces to sensory cells would convert their information, as it shows up, for example, in membrane ion channels, to electrical signals that could then be digitized and processed (DARPA MOLDICE 2003). Arrays of heat-sensitive cantilevers can sense infrared images without the requirement for cryogenic cooling; the reduced sensitivity could be more than compensated by cheap production and ease of use.

In many cases, the sensitivity of a sensing element decreases as its size shrinks – less radiation is collected on an antenna or light-sensitive surface, a pressure-sensitive membrane is deflected less, less nuclear radiation is absorbed in a thinner layer. In an acceleration sensor, a lower probe mass increases the resonance frequency and decreases the signal below that frequency. As a consequence, miniaturization to the nano- and even the micro-scale may in some cases be impractical.

In a sensor system – a combination of sensing element, processing electronics, communication channel and power supply – lower size limits may also follow from the requirements on the other components: a radio antenna may need to have a minimum area, the power source may have to be of a certain size. Nevertheless, NT will bring significant miniaturization potential with very small electronics and more efficient power sources.

For sensors embedded in another, larger system (such as a rifle, a transport pallet, an aircraft structure etc.) communication can be by wire and power can be provided by that system. Such sensors, fixed to a larger item, may be used in logistics or in monitoring the status of equipment. Sensors for environmental and security surveillance that are deployed at fixed locations, powered and communicating by wire, are somewhat similar.

The situation is different with distributed sensors for the battlefield. Here the sensor systems need to work autonomously, transported to arbitrary positions. Thus, a separate power supply is required and wireless communication by radio or by light will usually be chosen. Existing types of such sensors measure several centimetres (Blumrich 1998). With microelectronics and MST, systems could shrink to about 1 mm size, as foreseen in the earlier Smart Dust Project (DARPA MEMS 2003). Here, to solve the power problem, communication would be passive: a remote laser would illuminate the sensor which would open and close an optical shutter in front of a retroreflector, signalling by means of time pattern. This requires a clear line of sight. This is difficult to intercept, but exposes the carrier vehicle. However, if that is a cheap autonomous micro-aircraft, its loss may be acceptable.

A more flexible communication mode would be by radio via a self-configuring network of many scattered sensors. Transmission power requirements would be limited due to the low average distance to the next node, but of course significantly higher than with passive interrogation. Interception is of course easier here.

By using NT it is conceivable that distributed sensors for the battlefield reach actual dust-particle size, namely on the order of 100 μm or below. They could be produced at very small cost, allowing production in very high numbers. Hundreds or thousands would be scattered over an area of interest. Sensed quantities could include visible, infrared or radio-frequency radiation (including imaging), acoustic, seismic, magnetic signals, chemical or biological agents. The sensor networks would provide a more or less complete picture of the situation – locating and tracking enemy vehicles, monitoring enemy positions, locating enemy artillery, guiding own artillery and assessing its accuracy and effect, warning of chemical or biological attack etc.

Distributed sensors could also be used for co-operative verification of disarmament treaties and peace-keeping agreements – in the areas of conventional, nuclear, chemical and biological weapons. Here, however, small size and high numbers are not needed and would rather be counter-productive. Verification sensors would be deployed at few, agreed sites for a relatively long period of time; often even wireless transmission and autonomous power supply will not be needed. In many cases, traditional sensors of centimetres size would work equally well or even better than very small ones. NT could provide enhanced sensitivity and selectivity in detection of chemical and biological agents, but such sensing elements could be easily integrated.

Distributed sensors of the generic type making use of NT, connected to other pieces of equipment, could arrive within the next five years. The same holds for sensors for treaty verification. Very small distributed sensors for the battlefield will probably become available in 5–10 years.

4.1.11 Armour, protection

Effects on armour from NT will differ according to type. In heavy armour penetration is impeded by a thick layer of high-density material – mostly steel – sometimes augmented on the outer side by another layer or by explosive charges (active armour). What usually is advantageous with NT-based materials – much lower density – is that they would rather decrease the resistance against penetrating kinetic-energy projectiles or shaped charges. High-density metal will thus continue to be used. Some improvement could result from amorphous metal that is sometimes counted under

NT, even though it is produced by bulk processes (see Section 4.1.4). In the additional outer layer(s) and active armour, NT-based materials and explosives would bring some additional effect, too.

The situation is different in protection against smaller projectiles not designed for piercing armour. In light armour, NT-based materials such as composites from nanofibres, maybe with special structures, could lead to markedly better protection at reduced weight. Such light armour could be used in land and air vehicles.

Nanofibres could also be used to make stronger and lighter bullet-proof garments. Already a current DARPA program for personnel protection – not yet using NT – aims at 'ultra-lightweight armor with 100 percent improvement over current materials' (DARPA Budget 2002: 173; 2003: 198).

Layers with special nano-structures could provide specific absorption and reflection of electromagnetic waves and thus protect against high-power laser or microwave beams. (For protection against chemical or biological agents see Section 4.1.22.)

Limited improvement in heavy armour could arrive in 5–10 years. Significant change in light armour, including personnel protection, is also expected in 5–10 years.

4.1.12 Conventional weapons

Stronger and lighter materials would allow the building of conventional barrel-type weapons with reduced mass. Together with enhanced propellants, somewhat higher muzzle speed and accordingly longer range seem feasible. For ballistic and air-breathing missiles, the reduced mass could translate to a marked increase in speed, range or payload, and/or to a reduction of carrier size.

It is conceivable that *small arms* (for individual use) and *light weapons* (for use by a crew) could use barrels, locks etc. made of nanofibre composites, reducing the amount of metal/steel maybe even to zero. Also, metal-less ammunition could be made – the penetrating power of high-density material such as lead possibly replaced by a longer, thinner projectile of lower density. Depending on the difficulty of production and the strictness of controls, such light pistols, assault rifles, machine guns, mortars, rocket launchers etc. might be attractive for criminals, insurgents etc. Not easily seen by x-raying and without exciting an induction-loop metal detector, metal-less handheld firearms would pose strong problems for security checks.

Concerning *guidance*, NT would allow even smaller systems than MST alone. Inertial navigation systems, possibly augmented by satellite navigation (GPS), could be used not only in small missiles and artillery shells, but principally even in rifle ammunition and similarly small projectiles (for DARPA projects using MST-based inertial sensors, see DARPA MEMS 2003). Small control flaps could be moved by active material. With the higher accuracy from guidance systems, against some targets munitions of lower mass or smaller explosive charge could be used which would in turn allow smaller guns or carriers.

Kinetic-energy *armour-piercing projectiles* consist of high-density material, usually alloys of tungsten or (depleted) uranium, formed into a cylindrical rod with a sharpened tip.[6] The sub-calibre rod is accelerated in the gun barrel by a sabot that is discarded after leaving the muzzle. Alloying is required to increase the strength (Lanz *et al.* 2001). With densities of 19.3 and 19.0 Mg/m^3 for the pure metals, respectively, tungsten or uranium are at the natural limits of densely packing heavy atoms; NT could at most lead to small improvements of the alloy densities. However, the penetration capability is not only a function of the density; it is additionally influenced by the mechanical/hydrodynamic behaviour of the material. In depleted uranium, so-called adiabatic or localized shear helps to remove the deforming material at the penetrator head, sharpening it continuously. In attempts to achieve a similar process with tungsten heavy alloy penetrators, two concepts related to NT are being investigated. One relies on bulk amorphous metal, or metallic glasses.[7] Even though they are produced by bulk mixing of liquid metals and cooling, metallic glasses are sometimes presented under the NT rubric, probably because here properties of the different-size atoms lead to macroscopically observable elastic differences (e.g. IPSE 2003). Composites with nanocrystallites or nanowires would of course come under NT. First experiments with model penetrators made of amorphous metal showed indications of localized shear (Magness *et al.* 2001).[8]

The other approach to NT-improved penetrators is to start with nanocrystals of tungsten and the alloy metals, consolidating the mixed powder by hot isostatic pressing. Also here, indications for shear existed; perforating rolled heavy armour required 6 per cent less kinetic energy than with conventional tungsten heavy alloy. Both, amorphous metal as well as nanocrystalline tungsten alloy, show some promise of achieving similar penetration capabilities as uranium alloys, and thus could potentially replace depleted uranium 'with its perceived hazards and political difficulties'.[9] Future R&D may lead to improved values. However, it seems improbable that penetration capability can increase by more than about 20 per cent.

For the second principal possibility of penetrating armour, a *shaped charge*, metallic nanoparticles could be used in the liner, the conical layer that transforms to the slug (hot jet) on ignition (Burgess 2002: 10). Some improvement may also be possible by using NT-based explosive.

Conventional surface-to-surface *missiles* with ranges of hundreds of km have sizes above 10 m and carry on the order of 1,000 kg of payload, usually explosive. If the payload requirement remains, then NT-based propellant and structural materials could bring some reduction in total size and mass. Qualitative change would result if the improved targeting precision allowed the payload against point targets to be reduced to, say, below 100 kg – in that case, the missile could become correspondingly lighter, smaller (a few metres or below) and cheaper. The US Air Force Science and Technology Board even mentioned miniaturized 'intercontinental tactical ballistic missiles' (NRC Committee 2002: Ch. 6).

Air-breathing and rocket-powered missiles for shorter ranges (artillery, air defence) traditionally have sizes of one to several metres and carry dozens of kilograms of explosive. Also here, more accurate guidance could allow marked reduction of mass and size – provided that the target can be destroyed with the reduced energy transported. At present, small air defence systems that can be carried by a person (MANPADS) have missiles of 1.5 m length and 0.15 m diameter with a mass of around twenty kilograms. It is conceivable that a missile of less than one third of that size that directly hit a sensitive part of an aircraft could take the latter down.

NT with MST will allow much smaller missiles, down to a few millimetres diameter. The range would be limited to tens or hundreds of metres, and they could carry only grams of substance, delivering correspondingly low kinetic energy. But this could still suffice to kill a person – either by hitting sensitive parts of the body or by entering and setting off a small explosion that would lead to extensive loss of blood. Very small missiles could provide a recoil-less ammunition for metal-free, lighter small arms. It is unclear whether there will be military interest in such missiles (except maybe for special forces), but the potential is there.

NT will likely also play a role in new types of weapons. For example, electromagnetic acceleration of projectiles has been studied in several countries for decades (e.g. EML 2003). Different from traditional firearms, where the high-pressure hot gas driving the projectile cannot move faster than the speed of sound of that gas (around 2 km/s), there is no upper limit on muzzle velocity. Because air resistance increases in proportion to the square of speed, high-speed projectiles would be most effective in the high atmosphere and in outer space, but in the low atmosphere would also have higher kinetic energy and thus higher penetrating capability than traditional ones.

It is unclear if NT will overcome the obstacles to practical use that electromagnetic projectile acceleration has met up to now. Contributions are conceivable in electric power supply, including pulsed release (e.g. from flywheels), materials for barrel and projectile and of course guidance. NT might even allow using electromagnetic acceleration in small arms. No time estimate for eventual deployment can be given at present.

Beyond conventional weapons, NT contributions in power supply, materials etc. are likely in laser and microwave weapons as well as several types of non-lethal weapons.

Metal-less small arms and light weapons could arrive in 5–10 years. For NT-based small guidance systems probably the same time frame applies. Depending on the R&D results, improved armour-piercing penetrators or shaped charges can become operational in the next 5 or 5–10 years, small missiles rather in 5–10 years.

4.1.13 Soldier systems

NT can be used in various ways in systems close to, but outside of the body of the soldier, that interact with the body or enhance its functions. Sensors could measure various quantities – temperature, heart and breathing rate, transpiration. More sophisticated would be chemical sensors that detect, for example, stress-indicating molecules such as nitrogen monoxide (NO) in exhaled breath (such work is being done at the Institute for Soldier Nanotechnologies (ISN) (Talbot 2002)). Microneedles penetrating into the skin could be used to take samples of body fluids and inject drugs. Chemical and biological analyses could be done by integrated microfluidic biochips (DARPA BioFlips 2003). NT will provide opportunities for further miniaturization.

On the actuator side, there are possibilities for cooling or heating; NT- or MST-based pumps could circulate fluids through hollow (nano-)fibres in a garment. With appropriate valves, wound-healing agents could be applied where needed. Using controllably deforming material – e.g. electroactive polymers – parts of a battle suit could stiffen on demand to compress wounds or form a splint across a bone fracture. If the strength could be increased drastically, then an exoskeleton could become possible that would allow the lifting of heavier loads, or jumping farther than the unaugmented human body (for ISN work on polypyrrole see ISN 2 2002; Talbot 2002). Using NT-based filters, water could be reclaimed from body fluids and purified for re-use (see Section 3.1.6 and ARO Soldier 2001).

In limited amounts (average power microwatts to milliwatts), energy could be gained from body movement, e.g. by piezoelectric material. This could suffice in particular for the NT-based sensing and data processing parts with extremely low power demands. It is improbable that radio communication could be supplied from the same sources, since here power on the order of watts might be required. For thermal management with hundreds of watts it would certainly not suffice. Here, NT-based fuel cells might come in.

Much of the above, together with devices for data processing and communication, input and output, is contained in the battle-suit vision of the Institute for Soldier Nanotechnologies (see Section 3.1.6).

Whereas first external sensors for body functions can become available within five years, most will need 5–10 years. Drug-transporting and -releasing or stiffening material, thermal-management systems and integrated battle suits could arrive in 5–10 years, but a time above ten years is more probable.

4.1.14 Implanted systems, body manipulation

A further qualitative step in analysing, influencing and enhancing the soldier's body – including the brain and thus the mind – would use systems within the body. Since most of the body functions occur inside, in one sense this is the logical place for monitoring and manipulation. On the other hand, invasion of the body for other than medical reasons would represent a drastic change from present ethics and practice (for a short overview from a military standpoint on the principal possibilities that also mentions the ethical problems see Armstrong and Warner 2003; also NRC Committee 2002: Ch. 3).

Body functions to be sensed or controlled can be roughly categorized as biochemical – system-wide, at an organ or in a cell – or as neural, where the information content is the relevant quantity. Of course, there is no clear boundary here, e.g. the mood and cognitive performance are influenced by hormones, transmitters etc. A third category is mechanical enhancement.

In the first class, the body status can be sensed with a short response time from monitoring the blood chemistry. This applies, for example, to stress, injury, exposure to chemical or biological agents. Effects of the latter could be seen much earlier than the outbreak of external symptoms so that countermeasures would be more effective. On the actuator side, drugs, hormones etc. could be administered – on the one hand for therapy, on the other hand also in order to influence the mood – e.g. reduce anxiety, increase aggressiveness.

With continuing advance in understanding of biochemical cell processes, new options for controlling and manipulating them will appear. Goals of present US military R&D include inducing cellular stasis *in vivo* and *in vitro*, influencing metabolism to allow 3–5 days activity without intake of calories and the prevention of sleep-deprivation effects during seven days of continuous performance (see Section 3.1.2.3). While much of this and similar R&D will use biochemical methods, NT will contribute at least to the research methods, probably also to provide tools for targeting special sites in organs and cells.

A rather near-term NT application is connected with bio-compatible magnetic nanoparticles that selectively attach to bio-molecules or cells. They could be detected by small magnetic sensors and manipulated by magnetic tweezers. This could enable not only bio-detection and diagnostics, but also initiating and monitoring of functions within the cell, such as cell death, mitosis or protein expression (DARPA BioMagnetICs 2003). Future possibilities for manipulation in cells will likely allow marked modifications of human nature that could be used for many military purposes.

Concerning neural processes, microelectrodes for long-term contacts to neurones or sensory cells for measuring single-cell or collective excitation as well as for stimulation are routinely being used in medical therapy, others are tools in medical research. Examples of the former are cochlear implants, of the latter brain electrodes used with patients suffering from epilepsy or Parkinson's disease. Medical research is studying retinal implants for certain types of blindness and electrodes to sense motor-brain or -nerve signals and stimulate efferent nerves or muscles to overcome paralysis, e.g. from a spinal-cord lesion. Experience with patients and animal tests proves that reliable, long-term contact to nerves and brain is possible.

Electronic implants and microelectrodes already exist. NT will bring marked improvement in several ways:

- better biocompatible materials,
- smaller microelectrodes with improved contact to neurones,
- arrays with much higher numbers of microelectrodes,[10]
- smaller implanted electronics for signal processing and pattern evaluation with lower power consumption,
- power supply from body chemicals (see Section 4.1.5).

One can think of more futuristic concepts for NT-based nerve and brain contacts. One is to use insulated nanowires as electrodes that would be placed inside the brain by the blood stream in the arteries, avoiding opening of the skull and the *dura mater* (Llinas and Makarov 2003). Another is to let neurones and electrodes make contact by (directed, mutual?) growth.

In soldiers, nerve and brain electrodes could principally be used to gain signals connected with a simple intended action (moving a hand holding a control stick, flexing a finger to pull a trigger) and effect the action by technical systems faster. This would avoid some of the human reaction time (about 0.1 s) which may be important in certain combat situations (for the DARPA Brain Machine Interface program – that for humans envisages non-invasive signal uptake – see Section 3.1.2.3).

Implants could also be used for sensory enhancement. On the sensor side, widening the observed spectra of light and sound has been mentioned (ARO Soldier 2001). Beside seeing infrared and ultraviolet, or hearing infra- and ultrasound, one can also conceive of sensors for radioactivity or certain chemical or biological agents. Sensors would require some exposure to the outside, information would be transmitted by neural contacts. Subjects would probably need a considerable amount of learning and adapting. Because contacting a significant portion of the about one million neurones in the optical nerve will be very difficult, artificially imprinted images will probably have much less resolution than natural ones.

Implanted internal monitoring, data processing and communication systems – as already discussed at a US Army workshop, see Section 3.1.6 – would need nerve/brain contacts to transfer information. Transferring sensory impressions or simple concepts is conceivable using afferent sensory nerves. Communicating complex thoughts or feelings would need contacts to the associative brain cortex. Since the representation there is mostly unknown, major progress in brain research would be needed before such transfer could be envisaged. Of course, medical ethics strongly restricts human experiments for elucidating brain function.

Mechanical enhancement could work with 'integrated enhanced tissue, muscle, bones, tendons' (ARO Soldier 2001).

System-wide modification of biochemical processes in the body will probably arrive in 5–10 or 10–20 years. Targeted manipulation in body cells will likely become possible in 10–20 years. Implants for monitoring body status and releasing drugs could come into being in 5–10 years. Nerve/brain electrodes for signalling motion or simple sensory impressions would need at least 5–10 years. It is unclear when transfer of complex thoughts might become feasible. NT-enhanced tissue, bones etc. could arrive in 10–20 years.

4.1.15 Autonomous systems

Robots and autonomous vehicles have been the subject of military R&D for decades, but most have not yet been introduced into the armed forces – with the exception of pilotless small aircraft (drones, uninhabited air vehicles (UAVs)) mostly for surveillance; however, one type has been retrofitted with a missile.[11] Uninhabited combat air vehicles are beginning to be developed (see the project Advanced Aerospace Systems, DARPA Budget 2003: 213–225). While NT has not yet been used in projects for autonomous systems, its inclusion may finally lead to systems ready for actual deployment. NT may prove decisive via much improved computers, materials and propulsion, and could also contribute to weapons and ammunition. Detailed prospects for military robots cannot be presented here; the following is a sketch of the general possibilities.

Autonomous systems in the present context are all kinds of mobile systems that move without a person inside, are able to change speed and direction, react to conditions found in the local environment and have a size above about 0.5 m (for smaller systems see Section 4.1.16). This category includes robotic vehicles for motion on land, on and under water, in the air and in outer space. Propulsion could be by traditional means (wheels, tracks; jet engine), but also by unconventional ones such as legs or flapping wings. Included are all kinds of robots, independent of their shape – roughly human-like or not. Traditional target-seeking missiles, including cruise missiles, would marginally fulfil the criteria, but in most cases have been assigned a special target. Here, the focus is rather on systems that seek out targets that are not known beforehand. Of course, this would apply to target-seeking sub-munition, so some refinement of the definition beyond size may be needed to get a hold on the grey area that exists here.

Principally included are systems that can move only in restricted environments (such as in a factory, guided by induction loops in the ground), but the main emphasis is on systems capable of significant motion without auxiliary components pre-deployed in the environment.

In order to be inclusive, only motion autonomy is regarded here. Functional autonomy can come in several degrees; at one extreme there is full remote control all the time, at the other extreme the system would move and decide about its actions by itself after having been given general objectives. Intermediate degrees could use self-contained guidance on a small scale while the general course would be controlled externally, or could require remote control for certain actions (e.g. weapon release). Reasons for not including functional autonomy in the criteria are that it is very difficult to discern from the outside and that it can be changed fast.

A major sub-division of autonomous military systems depends on whether they are armed or not. Unarmed vehicles/robots could be used for tasks in logistics and transport, for security monitoring and battlefield surveillance, as radio relay stations, for search and rescue, for demining etc. Armed autonomous systems could carry or represent all kinds of weapons, from fire-arms via small and large missiles to nuclear bombs and chemical- or biological-agent dispensers.

On land, autonomous systems could resemble traditional cross-country vehicles, with or without armour, wheeled or tracked, including those with special functions such as missile launchers, artillery, radar, command and control etc. Removal of the crew may allow some reduction in size, but a heavy armoured vehicle with gun above 5 cm calibre will continue to have several ten tons of mass. New types of mobile systems could use legs and have shapes similar to humans or animals, maybe adapted for difficult terrain or urban environments. Others might be able to move under water on a lake or river bed (for DARPA programs, see Section 3.1.2.3).

For surface ships, those of larger sizes will probably continue to have crews, in particular those carrying nuclear weapons and aircraft carriers. Autonomy is mainly probable with small boats. Similarly, under-water submarines of larger sizes will likely not become autonomous. However, small submarines could work without a crew, e.g. for surveillance and shadowing, or could act as 'cruising' torpedoes.

In the air, all kinds of military fixed- and rotary-wing aircraft could fly without pilots. Special motives would exist with aircraft in combat situations. In particular, dispensing with life-support systems could make combat aircraft lighter, and maximum accelerations would no longer be limited by the pilot's endurance.[12] The higher agility could prove decisive, if the controlling program is intelligent enough or if the remote human operator reacts fast enough. The weight reduction, also deriving from NT-based materials and engines, would be most relevant for aircraft. Nevertheless, as long as about the same mass has to be carried in form of bombs, missiles, ammunition, the reduction in aircraft size and mass will be limited.

For stationary air vehicles such as blimps or permanently loitering planes, NT will provide improvements in hull membranes, structural materials, solar cells etc. Such aircraft at high altitudes (15–20 km) could be used for surveillance or as communication relays (for a project using hydrogen, probably not yet involving NT, see DARPA Budget 2003: 177). However, they would be very light and slow, thus highly vulnerable.

In outer space, autonomous craft could be used to service and repair one's own satellites – or dock at others' satellites and manipulate them. Here, NT-enabled mass and size reduction would be even more important than for aircraft.

NT-based autonomous systems on land, on and under water can probably be ready in 10–20 years; first autonomous aircraft using NT could arrive in 5–10 or 10–20 years. The latter time frame probably applies to autonomous spacecraft, too.

4.1.16 Mini-/micro-robots

NT would play to its full advantage in small and very small autonomous systems. Here, small (mini) means below about 0.5 m, very small (micro) below, say, 5 mm size. Such robots could move in all media, on land, in/on water, in air (for small robots in outer space where special laws of motion apply, see Section 4.1.18). Whereas with MST, micro-robots of centimetres, maybe a few millimetres size could be built,[13] NT will likely allow development of mobile autonomous systems below 0.1 mm, maybe down to 10 μm (this is still 2–3 orders of magnitude above the size range around 100 nm envisioned for nano-robots and universal molecular assemblers in MNT). Micro- and mini-robots in the size range 10 μm to 0.5 m would make use of NT in structural materials, energy storage and conversion, sensors, data-processing and actuators.

Small size brings principal problems: obviously the payload gets smaller. The energy supply will become more problematic the smaller the systems become. Smaller sensing elements collect less radiated energy or substance; for some quantities, e.g. chemical substances, increased sensitivity of nanoscale sensing elements may compensate this loss. Smaller communication antennae will suffer from lower signal strength; in some cases, the reduced ability to form a beam will also be a problem (due to diffraction, the width of a beam increases as the ratio of antenna size versus wavelength decreases). Smaller size impedes mobility, too, because viscous forces (friction) in the surrounding medium become relatively stronger. This is particularly relevant for very small systems in air and water (for a physics-based analysis of the motility of micro-robots see Solem 1994).

Limits in payload, sensitivity and communication range could be compensated by short distances – to the target and to other mini-/micro-robots – and by high numbers of systems. Limits in speed and range could be overcome by larger carriers that transport the small systems to the general area and release them there. Conceivable are aircraft, including crewless ones, missiles, artillery projectiles, but also land and water vehicles. Very small systems could be dispersed by wind.

Autonomous propulsion could use traditional technical principles: on land, wheels and tracks; on and in water, propellers; in air, propellers, jet engines or rotary wings. For smaller systems biomimetic or other principles might be preferred: on land, walking by insect-like legs, maybe with enhanced adhesion to master vertical or overhanging surfaces, or hopping, maybe using an explosively protruding rod; in water, undulatory body motion, a rotating or undulating flagellum, moving fins, on the floor (also for amphibious systems), walking legs like those of a lobster; in air, flapping wings (Bioderived and Bioinspired Materials: DARPA Budget 2003: 205–208). Biomimetic propulsion would be well adapted to muscle-like actuators.

A special case would be micro-robots that are small enough to move in the human body, e.g. for surgical operations, as in the vision of the 'micro submarine' in the blood stream unclogging arteries. Use in military medicine would probably be in parallel to civilian developments.

Mini-/micro-robots could be used for various military purposes. Functions for unarmed small robots include: surveillance and reconnaissance, path finding, sensing of chemical/biological agents, signals and communication intelligence, jamming from close distance, communication relay, deception (various programs: DARPA Budget 2003: 178–179, 186–187, 188, 189, 341–346). Acting as a target beacon, they would be integrated into a larger weapon system. Direct use as weapons is possible, too: small-scale actuation may lead to relevant damage if applied at a central node – a single small robot would seek out a target object or subject, approach and maybe enter it, move to a sensitive location and then cut a wire, explode a small charge or inject a toxic substance. For an unprotected human, the kinetic energy of impact by a small aerial vehicle could already be lethal.[14] Alternatively, mini-/micro-robots could be used in high numbers. Collectively, they could choke air intakes, block windows, put abrasives into mechanics etc.

Mini- and micro-robots acting in swarms, with continuous mutual communication and distributed intelligence, could become very powerful tools for applying force. However, developing the software for co-operation and goal-oriented behaviour adapting to actual circumstances is an extremely difficult endeavour. DARPA has several programs, among them one on swarms of mini underwater vehicles (Piranha: DARPA Budget 2003: 374–375, 410–411) and one on Software for Distributed Robotics (DARPA SDR 2003) (for others, see Section 3.1.2.3).

Mini-robots based on NT could arrive in 5–10 or 10–20 years. For micro-robots, rather 10–20 years will have to be reckoned – and the same holds for medical micro-robots for surgical operations in the body.

4.1.17 Bio-technical hybrids

Instead of fully artificial robots, one could use animals that have been manipulated to fulfil military tasks better than is possible with traditional training. Animals already provide the basic functions of energy uptake, storage and use, of locomotion and navigation, of sensing and processing, and of external manipulations, that may take a long time to produce fully artificially. Some additional components (sensors, communication devices) will be mounted outside of the body. Already using MST, additional sensors, nerve/brain contacts, control electronics and energy supply would be small enough for implantation into small mammals such as rats or mice, into birds or fish. Incorporating NT, similar systems would fit into small insects such as mosquitoes or fruit flies. Of course, the problem of economic routine implantation would have to be solved.

A first success was achieved with rats: microelectrodes for steering were implanted into the brain cortex where the left and right whiskers are represented; rats were trained by stimulating electrodes in a reward centre, as in the self-stimulation experiments of the 1950s (Talwar *et al.* 2002; DARPA SUNY 2003; see also Olds 1958). An experimenter could 'remotely control' the rat through complex environments in the laboratory. In military applications in the future, radio links could provide video from a camera on an animal and/or signals from animal sensors. Achieving greater distances would be possible with repeaters in the vicinity, e.g. on board mini-aircraft. Later, autonomous control by implanted computers is conceivable.

Research in how to utilize the functions of animals will also provide knowledge about how to implement intended functions in artificial robots, and to do so with minimal effort, e.g. by neural processing with small numbers of neurones at little energy consumption; with NT-based components, energy demand could become lower. Understanding how cells and organs develop for some specific functionality may later also provide methods to 'grow' most of the additional systems, which would be easiest if it could be done by genetic modification.

An alternative form of hybrid system would integrate animal organs into artificial systems; this seems attractive, for example, for the olfactory sensors of some insects which can already detect a few pheromone molecules, if they can be modified to molecules of military interest and continued functioning and reliable connection can be ensured.

At some point in NT development, the difference between living and artificial will probably blur, e.g. when biomolecular computers are connected to neurones, new traits are programmed genetically, or synthetic organisms are created.

Mobile bio-technical hybrids could be used for the same purposes as mini-/micro-robots, from reconnaissance to attack (see Section 4.1.16). Augmented animals would be well prepared for certain environments, e.g. birds or fish. While most species would provide some possibilities for small-scale actuation (using limbs, claws, the beak, the mouth, the whole body), some are equipped with organs that could immediately be used for attacking biological systems and humans, e.g., a stinging insect could inject a toxin. The lethal quantity for a human is below 50 μg for d-Tubocurarine (arrow poison curare), and below 1 μg for botulinum toxin (estimated with <100 kg body mass from LD50 values (e.g. Timbrell 1989: Ch. 9; Klaassen *et al.* 1986: 12)). These masses would fit respectively into 1/20 mm^3 and 1/1,000 mm^3, or cubes with 370 μm and 100 μm sides. These volumes and masses could be carried easily by a mosquito; for larger ones, wasps might be used. For biological/microbial (or hybrid/artificial) agents that self-replicate in the target organism, much smaller amounts would suffice so that even smaller insects could be used. The principle is demonstrated by malaria.

Because of the functional similarity and the blurring between living and artificial systems, bio-technical hybrids should be classed with mini-/micro-robots in considerations about preventive limitation.

First remotely controlled animals could 'enter military service' in 5–10 years. Hybrids with a higher degree of autonomy will probably take 10–20 years.

4.1.18 Small satellites and space launchers

Already MST will allow marked reduction of the sizes of satellites and space launchers (Altmann 2001: Section 4.2.6). NT will lead to even smaller systems due to size and weight savings in computers, sensors, structural materials, propellants, power supplies etc. (e.g. NRC Committee 2002: Ch. 6). Principally, satellites far below 10 g mass are conceivable, launched by rockets below 1 kg. To save energy loss due to air resistance, the small rockets could be launched from aircraft; with a payload on the order of 1,000 kg, many satellites could be launched with one flight. Alternatively, the rockets could get first-stage speed in electromagnetic launchers, preferably on high mountains. Traditional space launchers could carry high numbers of small satellites, maybe together with a large one.

Whether extremely small satellites will prove practical and can be produced economically is unclear. Sizes between 1 and 20 cm, however, should allow much lower launch costs per mass than today. Trajectory changes could be effected by ignition of very small explosive charges on the outer surface, as already developed with MST (since there is no air, control flaps and lift forces cannot be used to change the trajectory; only the rocket principle with expulsion of mass can be used).

Small satellites could work in swarms, with formations extending over 10 m to more than 1 km (e.g. NRC Committee 2002: Ch. 6). For radar, passive radiometry or communication the large aperture formed by the sparse array of receivers and transmitters would result in much higher target resolution than is possible with an antenna on board one large satellite. Since the combined antennae cover only an extremely small part of the virtual aperture, the received signal does not increase with the size of the latter. The transmitted power is likewise limited by the solar-cell area carried by the small satellites. Transferring satellite functions to swarms of small satellites would reduce the effects of a satellite failure and is one possible countermeasure against anti-satellite weapons of an opponent.

Small satellites are also foreseen for inspection of bigger satellites from close distance; this requires rendezvous capability. In another concept, they would dock and service or repair the other satellite. The same capability could also be used for anti-satellite weapons: foreign satellites could be approached and manipulated, rendering them inoperable or damaging them. Receivers could be jammed, solar cells covered, antennae screwed off etc.

Anti-satellite action could also be carried out by collision; due to the high relative speeds in outer space (several km/s), even small objects act destructively. To achieve a hit poses strong requirements on the sensing and guidance systems, but the principal feasibility of hitting a satellite has already been demonstrated without any NT. If hundreds of small collision satellites were available, the hit probability could be increased by using several of them on each target.

The US military have a few small-satellite R&D projects (DoD 2001: Ch. 5, 12; DARPA Budget 2003: 146–150, 215–220). The TechSat 21 experimental program of the Air Force Research Laboratory looks at clusters that function as a larger single satellite. DARPA is carrying out the Responsive Access, Small Cargo, Affordable Launch (RASCAL) program for rapid launch of satellites below 50 kg and the Satellite Protection and Warning (SPAWN) program for spacecraft inspection and space environment monitoring (for work on propulsion suitable for launch of 200 g satellites see Section 4.1.6). Whereas use of MST in manoeuvring miniature inspection spacecraft is explicitly mentioned, NT occurs only in general terms. There is no doubt, however, that advances from NT will be taken up fast (see also NRC Committee 2002).

NT-based small satellites and launchers could become operational in 5–10 years, more capable versions in 10–20 years.

4.1.19 Nuclear weapons

4.1.19.1 Auxiliary systems

Nuclear weapons have sophisticated systems for guidance, for safety and security, and for fusing. These are called Permissive Action Links (PAL) in the USA and incorporate sensors, mechanical locks, electrical and electronic circuits, etc. PALs and the simpler safety, arming and fusing devices for conventional weapons can be made smaller and lighter by MST (Altmann 2001: Section 4.2.4). NT will allow further miniaturization and probably better integration of sensors with mechanical, optical and electronic locking elements.

Some quantitative change is conceivable for the chemical explosive producing the critical fission mass – by implosion of a hollow sphere (pit) or by fast joining two sub-critical masses in a cannon-type assembly. NT-produced new bulk explosives with somewhat higher energy density (see Section 4.1.8) could allow a decrease in mass, or some increase in compression which would reduce the amount of fissile material required for ignition of the primary.

Principally, NT could allow the plutonium pits to be produced to even finer mechanical specifications than the extreme precision achieved already over decades of intense efforts. It is unclear whether this could lead to stronger spherical compression and thus lower critical mass, because it is well possible that the limits are rather given by non-sphericity of the detonation front in the chemical-explosive lenses. One can speculate whether explosive which is structured on the micro or nano level for a spherical detonation front, e.g. by gradual variation of composition – such work is being done at the US Livermore Laboratory in the context of the Stockpile Stewardship program (Parker 2000a, CMS 2002: 41; see also Section 3.1.4) – can lead to higher compression. Reductions of the critical mass by more than a factor 2 seem implausible, however.

All such improvement would not change the minimum requirement for a nuclear weapon, namely that it has to contain a few kilograms of fissile material. Together with chemical explosive, tamper, housing and auxiliary systems the weapon will at a minimum have a mass of several tens of kg, and hydrogen bombs with a fusion secondary will be correspondingly heavier. That is, the characteristics will not be qualitatively different from the weapons that exist already: the lightest known nuclear bombs, atomic-demolition munitions or backpack bombs, release below 1 kt of TNT equivalent ($=4.2 \cdot 10^{12}$ J) with about 70 kg mass (warhead proper: 27 kg). A nuclear artillery shell producing 0.1 kt TNT has 54 kg mass. Typical two-stage H bombs in missile re-entry vehicles (RV) release, for example, 335 kt TNT with less than 363 kg RV mass (Cochran *et al.* 1984: 311, 54, 75).

The first auxiliary systems for nuclear weapons based on NT could become available in the next five years, expanded use of NT will be possible in 5–10 years.

4.1.19.2 Computer modelling of nuclear weapons

When a nuclear weapon is ignited, extremely complicated physical processes occur in very short time. Mechanical, thermodynamic, nuclear, plasma, radiation and other effects are tightly coupled so that the equations that describe these processes cannot be solved analytically. Rather, numerical simulations have to be run on computers. Nuclear-weapon design laboratories have always used the fastest computers available at the time. Requirements increase strongly when the material is modelled in more dimensions. Over decades, actual nuclear test explosions were required to find out whether a new warhead design worked as planned. This is no longer possible since the Comprehensive Test Ban Treaty of 1996. With its conclusion, the nuclear-weapon states have strengthened their efforts for large computers to simulate explosions. Three-dimensional computations became accessible only a decade ago. Actual work aims at understanding detailed effects of corrosion and other anomalies in order to keep the nuclear-weapons stockpile from deteriorating. In the USA, simulations are validated against the data gained from more than 1,000 nuclear test explosions. With this expanded knowledge it appears feasible for the USA to modify existing warhead designs – e.g. for earth penetration with reduced yield – based only on computer modelling. This would be done with a conservative approach (for an overview see Zimmerman and Dorn 2002).

NT will increase computing capacities further by many orders of magnitude. Warhead modelling will be brought to a much higher level of sophistication than today. In principle, it is conceivable that radically new designs will be developed only by computer tests and then built. Whether armed forces would have sufficient trust in the performance certificates derived from the computer simulations is open. There might also be a strong motive for actual tests, leading to withdrawal from the Test Ban Treaty.

Judgements on this question and estimating a time period when such an option could become real would require more detailed knowledge on warhead physics and computer simulation. (It is possible that reliable estimates could be derived without access to secret information.)

4.1.19.3 Very small nuclear weapons?

The energy release per mass of nuclear energetic material is more than ten million times that of chemical explosives: full fission of 1 kg uranium-235 produces about 20 kt (twenty million kg) TNT equivalent, complete fusion of 1 kg lithium deuteride produces about 65 kt TNT equivalent. (In an actual explosion, only some fraction of the nuclei react, thus a weapon has to contain more nuclear material per energy that is to be released.) Over the decades of nuclear testing and weapons development, much sophistication has gone into improved compression of the fission material by the chemical explosion, and complex combinations of fission and fusion are being used to achieve the highest possible yield for a given mass of nuclear material.

Qualitatively different weapons would be possible if fusion could be triggered by some other means than a fission explosion which produces a minimum energy on the order of 1 kt TNT. In that case, arbitrarily small energy release could be achieved. The energy of such fourth-generation nuclear weapons (Gsponer and Hurni 2000: Ch. 4; Gsponer 2002)[15] could be much above the largest conventional weapons where typical values are 1 t (1,000 kg) of explosive for bombs, 500 kg for cruise missiles, and 20 kg for artillery shells – the upper limit is usually given by the capacity of the respective carrier/munition. Micro-fusion weapons, on the other hand, could release between 1 t and 1 kt TNT equivalent, while having masses of kilograms or tens of kilograms, principally even below a kilogram. They would provide new options for military attacks on dispersed or hard targets, and would blur the distinction between conventional weapons and weapons of mass destruction. Direct fusion ignition is feasible by several means; in inertial-confinement fusion research, one method is focusing many laser beams of extremely high energy on a very small (<1 mm) pellet of intricate structure. Such lasers fill large halls: the US NIF, for example, measures 200 m by 85 m.[16]

MST and NT could play a role in micro-fusion weapons (Gsponer and Hurni 2000: Ch. 4): In order to ignite fusion material, it has at first to be strongly compressed, and then heated even more at some spot; providing neutrons for the first fusion reactions can help. For these steps, laser or heavy-ion beams could be used, potentially augmented by neutrons from a subcritical fission reaction. Whereas miniaturization seems difficult for lasers and ion accelerators of sufficient energy, MST or NT could in principle become important if antimatter were used for compression and ignition. Antimatter annihilates when it meets normal matter, and releases the total energy $E = mc^2$ corresponding to the vanished mass m ($c = 3.10^8$ m/s is the speed of light): 1 kg of proton-antiproton mixture produces 22 Mt (22,000 kt) TNT equivalent. Because antimatter has to be produced in the first place at large particle accelerators such as CERN in Geneva at extremely slow rates, direct antimatter bombs are excluded. However, microgram quantities could suffice as a fusion trigger. Production of antiprotons and anti-hydrogen has been demonstrated, as well as storage over hours in special electromagnetic traps. Should one day MST or NT allow construction of micro-traps to safely keep a sufficient number of antiprotons or anti-hydrogen atoms from hitting the walls for a long time, and to let them reliably impinge on the trigger material when wanted, micro-fusion weapons of kg mass with energy release of tons of TNT equivalent could arrive. In principle, one can also not exclude alternative storage methods, such as metastable states, Cooper pairs, or distributed Bloch states of antiprotons in condensed matter. Should such methods be feasible, NT could help to achieve them. MST and NT could then allow cheap mass production.

Micro-fusion nuclear weapons enabled by MST or NT are not imminent. They can, however, not be excluded on theoretical grounds. Because of the extreme damaging effect they could have on international stability and peace, the field of antimatter storage and antimatter fusion ignition should be monitored systematically, as well as other means of fusion ignition with small devices.

4.1.20 Chemical weapons

Since NT can contribute to new chemical agents as well as to weaponization (mechanisms for storage, delivery, release, entry into the target body etc.), the general notion of chemical weapons is used here. Chemical weapons can be subdivided according to the target class. Herbicides act against plants (and are not subject to the Chemical Weapons Convention (CWC) (CWC 1993)).[17] The agents against humans (and animals) are either lethal or non-lethal. In the former class are agents that affect the lung, the blood, the skin or the nerves. Non-lethal chemical agents include irritants, hallucinogens, narcotics, convulsants, algogens (pain-evoking substances) and others (e.g. CBWCB 2003; on chemical weapons in general, see Evison *et al.* 2002 and http://www.opcw.org).

There is no doubt that NT can promote the discovery or creation of new agents in all these classes. Pharmaceutical research is looking to NT as a means to better understand mechanisms of toxicity, illness or pain and to design drugs that bind to specific sites in specific organs. Nanoparticles are foreseen as capsules for agents and as carriers across the blood-brain barrier.

These and other mechanisms could also be used to devise toxic agents and weapons. NT could be used in many forms, among them:

- capsules for safer enclosure and delayed release of agents,
- active groups for bonding to specific targets in organs or cells,

- vectors for easier entry into the body or cells, in particular in the brain,
- mechanisms for selective reaction with specific gene patterns or proteins,
- reducing the risk to one's own side by limiting the persistence or an improved binary principle.

NT (with biotechnology) will probably allow sophisticated interactions. These could include activation only within humans with certain DNA patterns, temporary incapacitation or influencing mood.

The USA is a party to the CWC. This is probably the reason why there are no public examples of chemical-weapons R&D related to NT.[18] The closest projects uncovered by a non-governmental watchdog organization concern microencapsulation of non-lethal chemical agents (such as binary agents or malodorants) and of bacteria (Sunshine 2002).

First new chemical weapons making use of NT could arrive in 5–10 years; sophisticated ones would probably take 10–20 years and more.

4.1.21 Biological weapons

Also here the general notion of biological weapons is used to include agents and weaponization components. Biological weapons use organisms (viruses, bacteria, rickettsiae, maybe also fungi and protozoa or even higher organisms such as insects) for hostile purposes. Toxins (toxic chemical agents produced by large or small organisms) are sometimes also counted as biological agents, but the Biological and Toxin Weapons Convention (BTWC) treats them as separate agents (see Section 6.4.7; the BTWC covers toxins independent of the production method) (BTWC 1972). The major difference to microorganisms is that the latter self-replicate in the target organism.

Together with biotechnology, NT will facilitate the creation of new microorganisms that could be used as biological weapons. Even today, improved understanding of the molecular basis of pathogenicity and better possibilities for genetic manipulation have been used to produce resistance against antibiotics, transfer virulence genes, suppress the immune system and create viruses (albeit simple ones) from scratch – most of this not for military purposes, and not yet using NT (Nixdorff *et al.* 2003: Section 11.3; synthesis of a bacteriophage genome: H.O. Smith *et al.* 2003).

NT will enhance the potential of biotechnology including the design of microorganisms with intended characteristics. The traditional properties of incubation period, epidemicity, infectivity, persistence, instability and retroactivity could all be optimized for military uses (for explanations of these terms see Nixdorff *et al.* 2003: Chs 2, 3). Similarly to chemical weapons (Section 4.1.20), new biological agents could make use of NT via

- capsules for safer enclosure and delayed release of agents,
- active groups for bonding to specific targets in organs or cells,
- mechanisms for easier entry into the body or cells, in particular in the brain,
- mechanisms for selective reaction with specific gene patterns or proteins,
- mechanisms to overcome the immune reaction of the target organism,
- reducing the risk to one's own side by limiting the persistence, programmed self-destruction, activation or deactivation by a second agent or reliable inoculation.

Some of this could be achieved without NT, using existing methods of genetic engineering and molecular biology. However, NT will provide many new possibilities in analysis and synthesis. Also with biological agents, sophisticated effects are conceivable – targeting of certain individuals, acting on the immune system, specific organs or brain centres (see also Petro *et al.* 2003). The feasibility of ethnic weapons is doubtful at present, since genetic variability is higher within than between groups. However, they cannot be ruled out for the future; also, a special population could be made vulnerable by genetic markers (Nixdorff *et al.* 2003: Section 11.3.1).

Development of new biological weapons would start with modifications of existing microorganisms, but at some time fully artificial systems might be used.[19] Different from chemical agents, the self-replication capability makes biological agents particularly dangerous. One single organism could suffice for infection as well as for (illegal) transfer. In addition, once released, uncontrolled evolution to unintended forms and interference with natural organisms, including gene transfer, cannot be excluded.

The USA is also a party to the BTWC. Thus, no NT-related biological-weapons R&D is expected except for defence (see Section 4.1.22).[20]

While the first new biological weapons based on biotechnology could be ready within the next five years, those making use of NT would probably arrive in 5–10 years, more sophisticated ones in 10–20 years or even thereafter.

4.1.22 Chemical/biological protection

NT provides possibilities of protection against chemical or biological weapons, for military as well as civilian installations and personnel. One class concerns earlier or more sensitive sensing so that other protective measures can be taken. A second type would block molecules by pores, a third one would degrade or destroy the agents, often by the large surface area of catalytic nanomaterials. The latter two could be used in filters for gas masks, air intakes etc., on protective suits or applied to contaminated areas after attack to neutralize the agents.

In the area of sensors for chemical or biological warfare agents, several NT-based approaches are conceivable. Specific binding to antibodies could be detected by fluorescent or magnetic nanoparticles or by changes in the resonance frequency of vibrating cantilevers.

DARPA programs in Biological Warfare Defense and in Biological Sciences look at a variety of principles; for some, NT may already be involved, or would lead to improved performance if introduced in the future (DARPA Activity 2003; DARPA Diagnostics 2003; DARPA Biosensor 2003; DARPA Genomics 2003; DARPA Countermeasures 2003; General: DARPA BIOS 2003):

- measurement of the reaction of living cells on chips,
- fast recognition of infection in clinical samples or in the body,
- improved biosensors, e.g. using enhanced or replaced antibodies,
- genomic sequencing of pathogens,
- unconventional countermeasures, e.g. robust vehicles for delivery of countermeasures into the body or modulation of the immune system,
- single-cell organisms or plants as ‘sentinels’ for specific substances by manipulation of their DNA so that the reaction changes an observable property, e.g. colour.

Chemical agents could change the electronic properties of conductive polymers, similarly to explosives (for activity at the ISN see Mullins 2002).

Concerning protection proper, NT could be used in porous membranes that would block all but the smallest molecules. Catalytic action of nanostructured material could absorb or destroy chemical or biological agents (for first tests see Klabunde 2000).

At the ISN, work is directed towards covering surfaces with reactive biocidal molecules, biocidal nanoparticles coated with dendrimer copolymers for destruction of chemical agents, and genetically engineered bacteriophages (viruses) that act as sensors as well as antidotes (Mullins 2002).

It is to be expected that many more types of NT-based sensors and protection equipment will be developed in the course of the increased R&D funding for US homeland security as well as in other countries.

The first NT-based sensors and neutralizing agents could arrive within the next five years, improved types in 5–10 years.

4.2 Summary of military NT applications

In a few military applications (explosives, heavy armour, armour piercing, nuclear weapons), NT will bring modest improvement. In many more, significant advance is foreseen, and for several areas the adjective revolutionary seems justified – either because of radical advance in existing applications such as electronics, computers, materials, maybe also software, or because of qualitatively new options, as with soldier systems, body manipulation, large and small autonomous systems, bio-technical hybrids, small satellites and new chemical/biological weapons.

Great efforts in R&D do not of course guarantee that the outcome will fulfil the original hopes. In some areas, the effectiveness or cost efficiency could turn out to be questionable. In particular, small systems could suffer from limited mobility, energy supply, communication capability or payload. Body manipulation might meet physiological or psychological problems. Artificial intelligence and autonomous systems could advance only slowly, as they did in the past.

The more generic military NT applications will have parallel civilian uses. Civilian R&D will be particularly active where mass markets are expected or where strong public interests exist – certainly with computers and software, some technologies of energy storage and conversion, medicine and maybe toy robots. In such areas, military applications will fast use civilian technology and may be driven by civilian R&D.

In the more specific military applications, however, where there will be little civilian demand or high technological risk, military R&D will certainly lead. In rare cases, costs may decrease after significant military investment and a civilian market may become possible. This might apply to mini-/micro-robots.

Table 4.1 summarizes the various military NT applications discussed in Section 4.1, together with the predicted times until possible deployment in four categories: 0–5 years, 5–10 years, 10–20 years, more than twenty years, and speculative. These predictions are rough estimates, and deviations are possible in both directions. Some indication is also given on the expected degree of advance.

4.3 Potential military uses of molecular NT

As discussed in Section 2.2, the feasibility of molecular NT (MNT) and its associated concepts is disputed, and there is no credible timing for their arrival. MNT proponents often assume a fast, even avalanche-like advance to the limits posed by the laws of nature. While that may be implausible, technologies that are at first sight independent of each other – such as universal molecular assemblers, nano-robots in the brain and superhuman artificial intelligence – would certainly accelerate each other. In conceiving of the potential military uses I assume that all the concepts of MNT proper as well as the wider ones associated with it will have been realized. This discussion is necessarily general, speculative and incomplete. The goal is to raise attention to potential problems that may arrive in the future, though possibly at different times. Much of this has been discussed in general terms by Gubrud (1997).

Table 4.1 Potential military NT applications, starting with more generic ones. The estimated time to potential introduction is designated by A (next five years), B (5–10 years from now), C (10–20 years), D (more than twenty years), u (unclear). Speculative applications where a time frame cannot be estimated are designated by '??'. The probable degree of advance is indicated in the right-hand column (+: modest, ++: significant, +++: radical).

Application	*Time*	*Features, Examples, Use*	*Advance*
Electronics, photonics, magnetics	A..B..C	Microelectronics, photonics, magnetics: smaller, more memory, faster, less power consumption; spin-/quantum electronics; small/light-weight/flexible displays	+++
Computers, communication	A..B..C..D	Very small, highly capable, with sensors, actuators, embedded everywhere; ubiquitous flexible networks; large-scale planning, decision, management; quantum computing?	+++
Software/artificial intelligence	A..B..C..D..??	Natural-language communication, translation, everyday knowledge; human-like interaction, perception, cognition, learning; autonomous decisions on many levels	++(+?)
Materials	B..C	Composites with nanoscale additives or carbon nanotubes; amorphous metal; self-assembled structures; various functional, active and smart materials	+++
Energy sources, energy storage	A..B..C	Fuel cells; solar cells; hydrogen storage; various small power generators	++
Propulsion	B..C	More efficient reciprocating/turbine engines; small engines; electric motors; biologically inspired systems (large/small); more efficient rocket engines (large/small)	++
Vehicles	B..C	Lighter, faster, more agile, longer range; electric propulsion	++
Propellants and Explosives	A..B	Better mixing of fuel and oxidizer, tailored energy density	+
Camouflage	B	Fast change of colour (pattern) according to background	++

Distributed Sensors			
Generic	A	Connected to pieces of equipment; surveillance at fixed locations – not autonomous	++
Battlefield	B	Sub-mm size, scattered in high numbers, interrogation by laser beam or self-configuring radio network	++
Verification	A	Centimetre size, NT only where special sensitivity	++
Armour, protection			
Heavy armour	B	Amorphous metal, additional outer layers/active armour	+
Light armour/ garments	B	Fibre composites, strong/light; nanostructures for absorption/reflection of electromagnetic radiation	++
Conventional weapons			
Metal-less arms	B	Nano-fibre composites in barrels, locks, munition	++
Small guidance	B	Inertial with MST, even in small projectiles	++
Armour piercing	A..B	Amorphous metal; nano-crystalline material; nanoparticles in shaped-charge explosive and liner	+
Small missiles	B	Size below 1 m against aircraft, a few mm against persons	++
Soldier systems	B..C	Outside body: sensors for body functions, actuators for release of drugs, thermal management, stiffening material, exoskeleton	++(+?)
Implanted systems, body manipulation	B..C u	Modification of biochemical processes; targeted manipulation within cells; implants for monitoring body status, release of drugs; implants for nerve/brain contact; enhanced tissue, bones etc.	++(+?)
Autonomous systems	B..C	Mobile on land, on/under water, in air, in outer space; traditional or new shapes/ modes of propulsion; unarmed or armed	++(+?)
Mini-/micro-robots	B..C	Mobile on land, on/under water, in air; various propulsion modes; unarmed or armed; special: surgical operations within body	++(+?)

continued

Table 4.1 Continued

Application	*Time*	*Features, Examples, Use*	*Advance*
Bio-technical hybrids	B..C	Small animals (rats, birds, fish, insects) with sensors, communication, nerve/brain contact for movement control; reconnaissance, small explosion, chem./biol. agent	++(+?)
Small satellites/space launchers	B..C	Monitoring and communication using swarms, inspection and servicing of satellites, fast launch; anti-satellite use by manipulation after docking or by collision	++(+?)
Nuclear weapons			
Auxiliary systems	A..B	Safety, arming, fusing	+
Computer modelling	u	New warhead designs developed without tests	+
Very small weapons	??	Hypothetical pure fusion without fission trigger, e.g. ignited by antimatter	?
Chemical weapons	B..C..D	Capsules, active groups for bonding, vectors for entry in body, cell, across blood-brain barrier, selective reaction, controlled persistence/reactivity	++(+?)
Biological weapons	B..C..D	Capsules, active groups for bonding, easier entry in body, cells, brain, selective reaction, overcome immune reaction, controlled activity	++(+?)
Chemical/biological protection/neutralization	A..B	Sensors, membranes, absorption, neutralization	++

In this scenario nano-robots move singly and in swarms, universal molecular assemblers self-replicate and produce all kinds of goods using mainly local resources, artificial intelligence has reached and transcended the human level and is carrying out R&D, testing and introduction of new technologies very fast automatically, processes in cells and the body are fully understood and controlled, human–artificial merging can occur in several forms, and outer space with its resources is being exploited on a wide scale.

4.3.1 Assembler-based military production

Fast-growing autonomous production at negligible cost could be used for more or less traditional military systems and weapons, but with all the advances achieved until then in 'plain' NT and other areas. Examples are rifles, guns, missiles, ammunition, sensor systems, communication systems, armoured vehicles, piloted aircraft, autonomous systems for monitoring or fighting. For applications in outer space, small and large satellites, launchers, kinetic energy as well as beam weapons etc. could be produced. (For nuclear weapons see Section 4.3.7.)

Production of equipment using traditional materials would be subject to limitations – the required raw material such as iron/steel, aluminium, titanium for structures, copper for wiring, would have to be brought to the facility. The same would hold for energy which could be transported in the form of electricity or as fuel (hydrocarbon, hydrogen). Some new types of weapons and systems would still need central installations, e.g. where big components are to be assembled from solutions containing special feedstock, or in vacuum. Also here, energy and raw material need to be brought. Other systems could conceivably rely (mostly) on locally available resources such as oxygen, nitrogen, carbon, silicon, calcium and other light elements that could be extracted nearly everywhere from the atmosphere, the soil or organic material in plants. Structural material, sensors, information-processing devices, actuators, energetic material: all would consist of appropriate arrangements of the various light atoms. The required energy for production could be gained from solar light and/or organic material. Within living systems, their metabolism could be tapped. Such decentralized production would favour small systems. Old and new types of chemical and biological agents should be accessible in this way, too.

Assuming that the production facilities for raw material, feedstock, energy and final products as well as the transport systems are themselves produced by MNT, a very fast increase of the production and distribution of military goods is possible – Gubrud (1997) even wrote of hours and days, assuming a doubling time of primary systems of fifteen minutes. In theory, a state could start with one universal molecular assembler and the corresponding assembly programs. As with every exponential growth in reality, limits of material resources, energy supply or transport would be reached ultimately, but until then enormous increases in military potential could be achieved. Whereas such a short-term scenario is not plausible for the introductory phase of MNT – when a complex process of learning and improving is to be expected, see Section 2.2.1 – it is conceivable for a later stage when MNT would be used widely. The process might be accelerated if it were directed by human-like or superior artificial intelligence. But even in the period of MNT introduction, a military emphasis could lead to a potential for fast growth at least of some weapons and systems.

4.3.2 Types of weapons

MNT could not only produce – and be used in – military systems that are similar to traditional ones, but would provide capabilities for qualitatively new means and methods of warfare. Exploitable MNT properties include:

- smallness,
- little resource consumption for production and operation,
- reliance on locally available resources,
- self replication,
- very high numbers,
- very high computing power,
- reasoning capability,
- sensors for all kinds of quantities,
- actuators from large to very small,
- mobility,
- capability to enter into objects and organisms.

MNT-based weapons could use various mechanisms to inflict damage: traditional ones such as explosion, heat release and kinetic-energy impact, radioactivity, electromagnetic-beam effects (thermal overload, mechanical impulse, electric overload), information attack against artificial systems or humans, chemical and biological interaction affecting matériel or organisms. Small weapons carriers could act singly, in swarms or in joint operations together with larger systems.

Small objects can carry only little payload. Important large targets (such as ships, bunkers, command-and-control centres, other infrastructure) will remain; physically damaging or destroying these will continue to require considerable energy release. Because tens or hundreds of kilograms of explosive will have to be transported, or penetrating projectiles of many kilograms will have to be shot, macroscopic carriers, guns or launchers would not vanish.[21] Another reason to keep large systems is that small ones have limited speed and range. They could be transported over long distances or against wind by pilotless aircraft or missiles that might measure only a metre or less. Without crews, armoured vehicles, aircraft carriers etc. could also become smaller.

What kind of mixture of macro-, micro- and nano-systems would evolve cannot be assessed at present. Principally, one will have to expect damaging systems on all size scales, from specially designed warfare molecules via nano-systems with and without self-replication, micro- and mini-robots to large weapon systems, some inhabited, many autonomous. They would move in all media, approaching their targets on and in the ground, on and in water and in the air. Given appropriate mechanisms for supply and transport, extensive underground excavation is conceivable. Outer space is a special case since here particular laws of motion hold. The following Sections 4.3.3 to 4.3.6 describe potential weapons mechanisms that could be used against military as well as civilian targets.

Many of the nano-systems would be able to self-replicate in an uncontrolled or a programmed way. This could take place before a target object/subject were reached, at it, or after penetrating into it. Following the MNT concept, self-replication could also be used to synthesize larger units on demand near or at the target.

As soon as military use of MNT came into view, the new systems would become main targets. Defending against an opponent's MNT-based systems and attacking them would become a major mission, and MNT would play a central role in that (see Section 4.4).

4.3.3 Operations against information and communication systems

MNT-based military systems could act against information and communication systems on the physical and the information level. Interference and damage can work in various ways; some are supported by smallness and the capability to enter or adhere:

- physical destruction of information channels from outside (antennae, sensors),
- penetration into carriers or command, control or communication centres with:
  - destruction of hardware – immediately or after settling down with delay (triggered by certain conditions – specified time period, external command, favourable state of carrier, e.g. in flight),
  - interference with information flow (stochastic or specific false signals applied to antennae, data wires; short-circuits, etc.),
  - taking control – certainly this would be extremely difficult.

False or misleading information could be fed into the system from a distance. Large-scale deception would be more successful, the higher the (artificial) intelligence that carried it out.

It is difficult to conceive of, but if one postulates a superior intelligence controlling one side, outsmarting an inferior opponent should be possible, in the 'optimal' case even without death, by cornering him through clever use of deception, overtaking command of sub-systems, denying resources etc. More probable seems a less clear-cut advantage with actual use of force, to the highest levels of violence. Independent of that the question is: why should such an intelligence remain obliged to the human command on its own side that may have initiated it earlier (see Section 2.2.2)?

4.3.4 Operations against equipment and infrastructure

Interference with, damage to or destruction of non-living targets (buildings, vehicles, mobile and immobile infrastructure etc.) could make use of

- traditional macroscopic effects from outside such as explosion, fire or penetrating projectile, with MNT support by guidance, material and artificial intelligence. MNT-based systems could also be used for (maybe distributed) transport to the target or on-site production;
- new systems of nano- or micro size acting from outside in masses: clogging openings for intake and exhaust, creating short-circuits, covering communication antennae, sensors or viewing windows, mechanically jamming hatches, wheels, legs, rudders, propellers etc., inducing wear by abrasives;
- small systems penetrating and acting on a small scale using traditional or new principles at sensitive, central locations: mechanical damage, thermal overload, electrical short-circuit, chemical etching – immediately or after settling in with delay.

These actions would be relatively easy for stationary targets. Moving land vehicles could be reached from the ground by hopping or adhesion, or by systems moving/waiting in the air. Because ships and submarines go relatively slowly and because nano-/micro systems could stay in the ocean without much energy expense, the former could also be attacked in motion. Due to their high speed, aircraft and ballistic missiles would mainly be affected on the ground and before launch. However, large swarms of very small air vehicles above important targets could thwart aircraft and maybe deflect ballistic missiles.

4.3.5 Operations in outer space

MNT could be used to enhance earlier types of space weapons, such as accelerators for projectiles, re-entry vehicles for terrestrial targets; laser, microwave or particle beams; or small satellites for impact or docking with subsequent manipulation, including better guidance and intelligent reasoning. The same holds for military launch capabilities and military infrastructure in space.

Of the several concepts of MNT-based utilization of outer space (e.g. Bishop 1997 and refs; Foresight 1995a; NSS 1995; McKendree 2001 and refs), most could be used for military purposes. 'Space towers' (carbon-nanotube-based ropes with elevators extending from equatorial positions 100,000 km into space, see Section 2.1.5) would enable more energy-efficient access to space; with them, military satellites of all sorts could be launched. Civilian exploitation of the moon, planets and asteroids by autonomous, self-reproducing systems for Earth would need accelerating the extracted material towards Earth, for asteroids changing their trajectory is envisaged. Such capabilities could also be used for military attacks by 'artificial meteors'. The material could also be made into weapons before arrival on Earth or on orbiting space stations. Concepts for improved civilian space use such as acceleration using laser or matter beams, self-healing satellite walls, intelligent lightweight space suits, repair of radiation-induced DNA damage in cells etc. would all be applied for military purposes as well. Other notions, such as interstellar travel and colonies in deep space, would be of little military interest, at least for Earth.

In outer space, high speeds (several km/s) are required to prevent falling down. Except for co-orbital rendezvous, target approaches occur at similar relative velocities and encounters last only milliseconds. Since in the vacuum of space there is no medium which could provide a lift force at a non-zero angle of attack, trajectory change cannot make use of rudders or control flaps. Expulsion of matter at high velocity is nearly the only method. For this reason one might doubt the efficiency of potential nanometre-size satellites. However, millimetre- or micrometre-size assemblies with solar cells and ion accelerators or photon sources for propulsion seem possible in principle.[22]

Because the only resource available in outer space is solar energy, self-replication and production of (military) goods in space is only conceivable by cannibalizing other satellites or by material gained from celestial bodies.

4.3.6 Operations against biological systems and humans

Living systems could be influenced, damaged or killed using MNT by a variety of mechanisms:

- traditional macroscopic action from outside – projectile, overpressure, heat – MNT-enhanced by guidance, material, intelligence, transport, near-site production;

- new action from outside by masses of small systems: clogging of nose and mouth, covering of eyes or face, mechanical blocking of fingers, arms, legs, creating slippery layers on the ground; for plants: clogging of cell openings, covering leaves;
- traditional or new chemical or biological agents, including application by injection by (artificial/hybrid) insect or release in nose;
- penetration into the human body with microscopic action, potentially after several generations of self-replication, starting immediately or after settling in with delay:
  - in organs: mechanical damage, clogging of arteries or inducing their contraction;
  - in cells: changing of metabolism or DNA, arbitrary results (assuming full understanding of gene and protein chemistry), e.g. new hair colour, over-/underproduction of hormones, excrescence in the face, tumours, necrosis of limbs;
  - in nerve cells and brain:
    - interference with perception, thinking or motor activity,
    - control of general mood, pain sensation etc.,
    - control of thinking (certainly extremely difficult to achieve).

4.3.7 MNT and weapons of mass destruction

4.3.7.1 Chemical and biological weapons

Chemical and biological warfare agents could of course be produced by MNT. Some new types could act very selectively and would thus not count as weapons of mass destruction. However, non-selectivity could also be designed – or it could result from errors in production or from modifications occurring after release, possibly by uncontrolled evolution. Chemical and biological agents of many types would be amenable to production by nanoscale entities that self-replicate in the wild or in the affected bodies. Virulence properties such as infectivity, epidemicity, incubation period, stability, retroactivity could all be designed, beyond those of known biological weapons (for explanations of these terms see Nixdorff *et al.* 2003: Chs 2, 3). Contagion could use many pathways, some known from existing diseases (such as inhalation, insect stings), some new (e.g. eating affected plants). By exploiting some feature unknown to an enemy, a large proportion of its population could be killed very fast, while one's own citizens could be protected through some form of vaccination, perhaps even unnoticed through the drinking water.

One can also conceive of chemical/biological agents that would not kill, but mentally disturb the target subjects. If a large enough number were affected for a sufficiently long period, the health system and society at large could break down, which could then indirectly lead to massive loss of life. Another, though long-term effect, would work via sterilization of the target population.

Finally, the threat of grey goo (general ecocide by omnivorous replicators) could be used by some states as a general deterrent, similarly to the 1950s idea of global radioactive contamination by a cobalt-60 bomb.

4.3.7.2 Nuclear weapons and nuclear deterrence

In the production process of nuclear weapons, all steps could be automated and much accelerated by MNT, from uranium mining via building and running of enrichment plants to plutonium-production reactors and reprocessing plants. Interesting questions are whether MNT would allow the extraction of uranium from seawater or soil, and whether it might provide a more efficient means for uranium enrichment – maybe by nanomachines that weigh and sort individual UF_6 molecules? (Uranium hexafluoride is the gaseous component that is used in enrichment by diffusion and centrifuge.) Certainly the bomb assemblies with casings, neutron reflectors, explosive lenses etc. could be made. For the plutonium pits, high-temperature casting may still be required, but the ovens, moulds etc. could be made by MNT processes.

Since there is no scarcity of uranium, small nuclear powers should be able to produce tens of thousands of warheads, the USA hundreds of thousands to millions over a period of years (Gubrud 1997).

Should very small nuclear weapons become possible, MNT could make them a cheap, mass-produced item; however, the antimatter required would still have to be produced by nuclear reactions in high-energy accelerators (see Section 4.1.19.3).

Nuclear-weapon use by states has up to now been (more or less) reliably prevented by deterrence: the near certainty that a second strike following a first attack cannot be avoided. MNT could conceivably change that and make victory in nuclear war (seem) achievable. Several mechanisms could contribute to such a development (Gubrud 1997):

- disarming attacks on bombers, land-based missiles and command-control systems could be launched, e.g. by swarms of nano- or micro-robots;
- vast numbers of very small submarine sensing systems distributed throughout the oceans could trace the nuclear submarines, making them vulnerable; torpedoes would not be the only method of attack – blocking the missile hatches could disarm them;
- active defence by massed interceptors of various types could significantly reduce a limited missile attack;
- deep underground shelters could be excavated and built, with closed-cycle life-support systems and a transportation infrastructure for fast evacuation of urban populations, greatly reducing the death toll.

One can have doubts on the plausibility of the last two mechanisms. Anyway, nuclear-weapon states will have strong motives to not let their second-strike capability become endangered. They could increase the numbers of their warheads, disperse them further and take measures to negate the MNT capacities of an opponent.

4.3.8 Protection capabilities of military MNT

In parallel to unprecedented new options for selective or general damage and destruction, MNT would also provide new possibilities to protect civilians and soldiers from attack, in particular by MNT weapons. This area is more uncertain than that of weapons, thus the following considerations are even more speculative.

Against macroscopic effects, MNT could provide somewhat better protection than earlier forms of NT. Information systems would have capabilities to monitor themselves. (Super-)human artificial intelligence would be used to detect all kinds of potential attacks, physical and informational.

Protection against nanoscale entities entering the body could principally be achieved by closed environments with corresponding filters for entering matter. Whereas this is conceivable for air and water, it seems impossible for food and other goods. Enclosing individuals by whole-body suits may be acceptable for soldiers in combat, but not for everybody all the time. Because complete enclosure of whole regions or countries will probably remain impractical, there would be motives to put one part of the protection into the general environment, another one inside the body.

In the environment, fleets of 'guardian' nano-robots, connected to monitoring centres, could patrol ground, water and air, checking for the presence of dangerous or unknown substances/entities. Since this would have to include objects down to molecule size, the practicality of such a scheme is unclear.

Inside the body, an artificial 'immune system' of dispersed nano-robots, with or without central co-ordination, could monitor the blood and the cells for unknown objects and get rid of them – e.g. by disassembling them or by enclosing and removing them.

Concerning nuclear attack, effects of overpressure and heat could not be negated except for the traditional means of distance and sheltering. Radiation effects on the body, on the other hand, could conceivably be reduced – at least up to a certain dose level – by nano-robots in cells. They would have made additional copies of the DNA for increased redundancy and would repair DNA damage more reliably than the biological mechanisms. (This was proposed in the context of preventing ageing and illness by Drexler (1986): Ch. 7.) Similarly, cell biochemistry could be monitored and corrected.

Even though in the context of a military MNT build-up these and other protection measures would be taken, there are good reasons to question their effectiveness:

- Self-monitoring of information systems could be switched off or disrupted by information attack.
- Artificial intelligence of (super-)human capability could be reprogrammed by information attack, could be 'convinced' to change sides or could redirect its goals on its own.
- Intruding systems could be disguised as 'normal' object/substance or as guardians.
- Guardian systems could be re-programmed for attack by an opponent; something could go wrong with them; released to self-replicate, they could evolve in an uncontrolled way.

Protection schemes as conceived above would constitute major interventions in the environment and the human body. Survival would be highly dependent on the behaviour of autonomous actors and unmonitored interactions. Events would be extremely complex and not transparent. Radical change could occur in a short time. Even superhuman intelligence might not be able to monitor and control what would be going on. Comparing the capabilities for damage with those for protection, it seems plausible that the old experience – it is easier to destroy than to protect – would remain valid.

4.4 Countermeasures against military NT systems

Assuming widespread NT application in the military as mentioned in Section 4.1, defence and offence against hostile NT-based systems attain central importance. This holds in particular for new systems such as mini-/micro-robots or chemical/biological agents. It would hold all the more if molecular NT became possible. Because of the wide variety of effects, various countermeasures are to be expected that would make massive use of NT themselves, of course at the respective technological level available. Some methods of countering NT-based weapons could be:

1. general strategies:
   - faster information processing, more autonomous decisions,
   - withdrawal of humans, use of mostly artificial systems,
   - redundancy – increase number of own systems,
   - dispersal of functions to many smaller systems,
   - hide better by smallness and camouflage;
2. passive protection:
   - observe the environment, get out of the way,

- sieves with molecule-size pores against penetration through openings,
- complete encapsulation, also of sub-systems,
- make adhesion to surfaces more difficult,
- hardening (e.g. nanolayer against mechanical abrasion or heat, stronger structures against impact);

3 active defence:
- small missiles/projectiles against approaching mini-/micro-vehicles,
- active surface for destruction of approaching or adhering objects,
- micro-/nano-robots as 'guards' (on outer surface, inside own systems/positions; within body/cells),
- preventive 'inoculation';

4 offence strategies (using various means):
- counterattack,
- pre-emptive attack,
- preventive attack.

The effectiveness of weapons vs. countermeasures is unclear at present. Similarly, one cannot predict which mixture of defensive and offensive means and methods may develop. However, there are no indications of defence dominance, so that counterattack and pre-emptive or even preventive attack will likely continue to play an important role in armed conflict.

5

PREVENTIVE ARMS CONTROL

Concept and design

This chapter describes the approach of preventive arms control. General considerations on limiting technology are presented in Section 5.1. Section 5.2 describes the process and a set of criteria of preventive arms control. Section 5.3 discusses the considerations that influence the design of preventive limitations.

5.1 General considerations on preventive technology limits

5.1.1 Technology limits in the civilian sector

Often one hears the argument that technological progress cannot and should not be limited. In such a debate, it is useful to remember that in civilian society there exists a body of regulations covering research and development (R&D) as well as the introduction of new goods into the market. Unrestricted use of scientific or technical advances is not always allowed: legal obligations affecting R&D have to do with protection of workers and the public against dangerous substances. There are licensing procedures for handling certain substances and limits for the concentrations at the workplace or for the amounts released to the air or to waste water, e.g. in the chemical industry or nuclear facilities. Specific security features are demanded for laboratories handling infectious micro-organisms or manipulating genes. While such rules mainly affect how R&D is being carried out, there are also a few areas where the substance of research is affected. One area has to do with experiments on humans which are only allowed in strictly circumscribed conditions. Another obvious area is research that would pollute the environment. A final, specific example is that in Germany the penal code forbids setting off a nuclear explosion – so all kinds of research that could be done with such explosions are prohibited (par. 307 of the German *Strafgesetzbuch*; perpetrators face up to life imprisonment).

After R&D, the process of bringing a new product to the marketplace is also regulated. New therapeutic drugs have to pass a complex testing and certification process. Safety standards have to be adhered to, etc.[1]

In recent decades, the insight has grown that in dealing with new technologies, there are sometimes unintended and unforeseen consequences. Important examples are DDT damaging fertility, asbestos leading to cancer, carbon dioxide from the burning of fossil fuels increasing the atmospheric greenhouse effect or the chloro-fluorocarbons that destroy the stratospheric ozone layer. In such cases, scientific proof of the consequences may only be possible after irreversible damage has already occurred. Thus, the philosophy on regulation of new technology needs to be changed to the precautionary principle. This principle holds that limits should not have to wait for full indisputable evidence, but should already be applied if there are reasonable grounds for concerns about potential dangers to environmental, human, animal or plant health (for the European-Union approach to the precautionary principle, see CEC 2000).

Thus, in civilian society R&D as well as the introduction of new products take place within a regulatory framework defined by the society, and the freedom of research is not unlimited. Within societies, the state monopoly of legal force and the judicial prosecution of perpetrators ensure that the rules are being followed to a sufficient extent. Citizens usually can rely on that process and need not prepare for all kinds of accidents and misuse of new technology by themselves.[2]

5.1.2 Technology limits in the military sector

In the international system between countries, the situation is markedly different. There is no overarching authority with a monopoly of legitimate force, and there is no penal code with criminal prosecution of perpetrators. Only first steps in that direction are being taken with the establishment in 2002 of the International Criminal Court that is to deal with 'the most serious crimes of concern to the international community as a whole', such as genocide, crimes against humanity and war crimes (http://www.icc-cpi.int; unfortunately the USA is acting against the ICC (e.g. HRW 2003)). Of course the UN Charter sets important norms against threats to peace and acts of aggression, however its mechanisms for dealing with these (Chapter VII) have not yet been enacted. Moreover, these mechanisms do not represent a higher authority with an actual monopoly of legitimate force. The UN Charter represents important progress from the earlier state when countries could legally go to war at will anytime. However, its Article 51 (right of self-defence until the UN Security Council has taken measures) is a clear recognition of the fact that the UN system cannot (yet) guarantee the security of its members.

This means that states continue to find it necessary to maintain armed forces for their security. Readiness to use military force for defence is usually linked with capabilities for offence, so that preparations for one's own security regularly increase threats to other countries and decrease their security. This is the so-called security dilemma. Its mechanisms are at work wherever armed forces exist and armed conflict seems principally possible – even if the corresponding states are partners in many respects. Ways out of the security dilemma exist: one is creation of political and economic (and maybe military) relations so close that armed conflict is practically excluded and unthinkable. The European Union provides an example. Another way out is mutually agreed limitation and reduction of armaments with adequate verification, as with the nuclear arms-control treaties of the Cold War and the Treaty on Conventional Armed Forces in Europe. A third possibility is to separate the defensive from the offensive function of the military – as far as that is possible – and strengthen defence while limiting offence so that clear defence superiority is achieved. (The conceptual problem here is how to make sure that a single defender is superior to a coalition of many attackers.) This strategy was traditionally used, for example, by Switzerland; first discussions about adopting it between NATO and Warsaw Treaty Organization (WTO) were stopped when the WTO dissolved. In principle it could provide a useful guideline for the military relations between USA and Russia or USA and China, however only if ambitions for world-wide force-projection capabilities were reduced markedly.[3]

With respect to new military technology, the security dilemma provides strong motives for fast innovation, because technological superiority is a decisive factor in armed conflict. When potential opponents are advancing military technology fast, often the consequence is that the insecurity grows for all sides. This may be due to weapons that can attack more targets, have longer range or better accuracy or that travel faster and demand shorter reaction times. The nuclear arms race of the Cold War provides many examples, most prominent the transition from fission to fusion (hydrogen) bombs, the introduction of the long-range ballistic missile and of multiple independently targetable re-entry vehicles. There are, however, also a few cases where insight led to preventive limits that blocked this mechanism – the Antiballistic Missile (ABM) Treaty (1972–2002) is a prominent example.

As arms-control treaties in general, preventive limitation of military technologies takes place as a voluntary act, agreed between sovereign states. Thus, the latter must see such limitation as being in their best interest. Since there is no higher authority that could enforce compliance, the states need to consider what to do in case a partner does not comply. One principal possibility is to covertly break the rules in advance in order to be prepared – however, this would damage the very purpose of the treaty, of course. The alternative is verification that allows to adequately convince oneself that the treaty partner is complying with the stipulations. In this case, a violation would be detected in time so that a reaction is possible and a military advantage is denied. The reaction could take various forms – politically one could talk to the partner and convince him to change his behaviour; if that does not help, a military reaction could be to build up compensating capabilities, which would usually require withdrawal from the treaty.

For preventing deterioration of their mutual security, potential opponents should agree on adequately verified limitation of those new military technologies that create strong dangers, in parallel to quantitative and qualitative limits on the already existing armaments. Ideally, such limitations could then, in the further process, be linked to a defensive restructuring of the armed forces at large. Continuing progress in threat reduction could lead to improvements of the political and economic relations that make armed conflict unthinkable.

However, it should be stressed that agreed limitation of new military technology as well as of existing arms and forces makes sense even if further progress seems impossible.

5.1.3 Technology limits and the tasks of the armed forces

Agreeing on technology limits means restricting the future fighting efficiency of one's own armed forces. The degree to which political decision-makers will agree to such limits will depend on many factors. One of them is the size of the effect; if a future technology is far from being used, it is of little military relevance, and thus limiting this technology does not actually restrain one's own capabilities. Such mutual agreement is not useless, however, because over time the technology may evolve to become militarily effective – and then the limit may prove effective finally. Alternatively, one can argue that agreement on topics of little relevance may build trust and help a political process that could later lead to substantial limits. As examples of this approach one could cite the Antarctica or Moon Treaties where the prohibition of military installations etc. is still not a substantial restriction of military capabilities, or the concept of breaking the present deadlock with respect to a comprehensive ban on space weapons by an agreement banning weapons beyond geosynchronous orbit (35,800 km altitude if circular) (Singer and Sands 2002). The problem here is, of course, that the agreement may not hold when the respective technology becomes militarily effective.

Factors strongly influencing decisions on limiting military technology are the tasks of the armed forces and the planned circumstances of their use. The tasks where armed violence would occur can be roughly classified as follows:

- large-scale armed conflict outside of one's own territory – against a superior, about equivalent or inferior opponent,

- defence of one's own territory against large-scale attack,
- crisis intervention, peace enforcement – globally or in the region,
- defence/protection against terrorist attacks.

It is obvious that the requirements on the strength of the armed forces differ widely between these tasks. Similarly, the need for fast technological advances is different. Preventive limits on new military technology should be the easier the lower one's military goals are ranked in the list, at least concerning offensive means. But even in the case of the strongest requirements, there remain good arguments for preventive limitation of technologies that increase threats and decrease stability. Unless war is seen as unavoidable, the armed forces are only being kept as a means of last resort, for being prepared in case war occurs. Preparations for that case should not make it more probable.

Obviously there is a tension between military preparations and arms limitations, preventive or not. Mutual limits will usually contribute more to international security than strengthened armed forces, but following this insight more often than not meets resistance within countries.

5.1.4 Preventive arms control after the Cold War

After the dissolution of the Soviet Union, the USA and Russia treat each other increasingly as partners, and relations between the USA and China have also improved. Even though armed conflict between them seems remote at present, nevertheless these states keep their respective forces and train them for such a contingency, among others. At the fundamental level, nuclear deterrence is still at work, too (also including the other nuclear-weapon states) (e.g. Steinbruner and Lewis 2002; Russia 2000, 2000a).

Thus, motives to use new technology in order to prevail in such a war – or to prevent an opponent from prevailing – continue to exist. As a consequence, a competition in military technology is in some way going on all the time. With the outlook for revolutionary change in many fields, NT could intensify this process drastically – accelerating arms races could develop in all areas of military NT applications (see Chapter 4).

One can even argue that to some extent, a similar mechanism is in effect between partners in a military alliance. On the one hand, there is the principal possibility that the political situation in the partner country may change at some time in the future, so that armed conflict and military threats can no longer be excluded completely.[4] On the other hand, arms-race pressures on allies work via the common potential opponent(s). And for the less fast advancing partners there is the argument that for co-operation in armed conflict, standardization and interoperability require preventing too large a gap to a technologically leading partner.[5] Thus, it is not difficult to conceive of a rush towards, for example, autonomous combat vehicles, should one country start introducing them, not only between potential opponents, but also among partners – and the same for micro-robots, mini-missiles etc.

This discussion makes clear that there are good reasons for preventive arms control also after the Cold War. The reduced mutual threat would allow in principle to agree on much wider-ranging limitations than were possible in former times.

A counter-argument could be that the strongest security threats at present do not stem from the states with strong armed forces, but from terrorist groups and failed states – and these could not be partners in preventive limitation. However, such groups and states are unlikely to be able to develop NT-based new weaponry by themselves. The much more likely scenario is that military technology and weapons developed in the high-technology countries will be exported or otherwise proliferate to end up in the hands of non-state actors. As a consequence, limits agreed among the NT-capable countries will have a significant effect in limiting access of terrorists and groups in failed states, in particular if preventive limitation is being supplemented by special efforts to limit dual-use exports.[6] Not providing qualitatively new tools for terrorist attacks or asymmetric warfare should create strong motives for limiting action.

5.2 Preventive arms control: process and criteria

The general concept of preventive arms control dates back to the Cold War, where one can find many instances when the military situation became more unstable and dangerous after new military technologies were introduced, such as the hydrogen bomb, the long-range ballistic missile or multiple warheads on nuclear missiles. After deployment, agreement on reduction was very difficult to reach. The goal of preventive arms control is to avoid similar situations and to prevent new military technologies with potentially dangerous consequences from being realized in the first place, before they are deployed (Neuneck and Mölling 2001; studies written in the context of a project of the Office of Technology Assessment of the German Parliament TAB include: Petermann *et al.* 1997; Brauch *et al.* 1997; Neuneck and Mutz 2000).

Preventive arms control consists of four steps:

1 prospective *scientific analysis* of the technology in question;
2 prospective analysis of the *military-operational aspects*;
3 assessment of both under the *criteria of preventive arms control*;
4 devising possible *limits* and *verification* methods.

These steps are to be carried out in interdisciplinary research, interacting with practitioners. In the optimum case, nations would afterwards start negotiating the corresponding agreement.

In the course of the German joint projects on preventive arms control, of which this work and its predecessor on microsystems technology (MST) (Altmann 2001) form parts, a refined set of criteria has been developed for finding out where military-relevant technologies may entail special dangers so that considerations on preventive limits should take place. These criteria are put in three groups (Neuneck and Mölling 2001):

I Adherence to and further development of effective arms control, disarmament and international law
- Prevent dangers to existing or intended arms-control and disarmament treaties
- Observe existing norms of humanitarian law
- No utility for weapons of mass destruction

II Maintain and improve stability
- Prevent destabilization of the military situation
- Prevent technological arms race
- Prevent horizontal or vertical proliferation/diffusion of military-relevant technologies, substances or knowledge

III Protect humans, environment and society
- Prevent dangers to humans
- Prevent dangers to environment and sustainable development
- Prevent dangers to the development of societal and political systems
- Prevent dangers to the societal infrastructure.

These criteria are not carved in stone, of course;[7] further debate and development are needed.

This set will be applied to the potential military NT applications in Chapter 6.

5.3 Design of preventive limitations

In principle, rules preventively limiting military applications of new technologies could take many forms. What forms are chosen will depend on general considerations, properties of the technology in question and political circumstances. Questions to be decided include: How to balance the various benefits, risks and costs? What is the appropriate degree of specificity for limitations? At which stage in the technology production process can and should limits be applied? How should civilian R&D be dealt with?

When these questions have been answered, one should strive for clear definitions with few problems due to a grey area between allowed and prohibited activities. Effort and intrusiveness of verification should be limited, and the whole package should be politically acceptable. Obviously, these considerations are complex, and proposals cannot be gained by purely scientific methods.

5.3.1 Balancing of benefits, risks and costs

When consideration of the criteria has provided reasons for limiting military uses of a certain technology, there will usually also be motives to use it for other, positive purposes, in particular in the civilian sector. Ideally, one would want to fully prohibit the dangerous military applications while not at all impeding the desirable uses. Such a clear-cut separation is often difficult, however, and the issue of how to draw the line(s) is intertwined with the issue of verification. Finding an appropriate solution requires pondering the advantages and disadvantages in several fields, and will often need a discussion of several options. (The effort for such analysis is nevertheless limited since it has only to be spent for those applications where limits are needed.)

If a specific technology is not to be generally prohibited, specific limits on it or its application usually rely on some parameters, such as size, power, speed or payload. Because in most cases there is some overlap between the negative and the positive uses, a cut at certain parameter thresholds will still allow some of the uses that one wants to exclude while prohibiting some uses that one would want to keep. In order to find the optimum or at least an acceptable demarcation, one has to balance the advantage (security gain) provided by the military limits with the disadvantage (loss in potential positive uses) of the corresponding civilian bounds. The fact that too wide bounds for the allowed activities would alleviate circumvention of the military prohibitions particularly needs to be taken into account. Fortunately, the parameter set available contains many dimensions that can be used to maximize the advantages and minimize the disadvantages. Additional tools are operational rules and verification. An example is provided by the graduated limits on high-power lasers that I have proposed for a ban on laser weapons.[8]

Concerning verification, there are specific costs and benefits (for general principles of verification, see Scheffran 1985, 1986; for guidelines on verification of R&D limits, see Altmann 1994). More intense verification reduces the possibility and likelihood of non-compliance and thus increases the security gain. However, at the same time the effort, financial burden and intrusiveness of verification grow. In case of national technical means, the effort and cost arise on the side of the verifying state; for co-operative verification, using on-site inspections and on-site equipment, there is also a burden on the verified country. Intrusiveness concerns privacy, including a danger to legitimate – commercial or military – secrets, and the impediment of normal work that inspections may entail.

The requirements on verification increase as the objects or activities to be monitored become smaller, more numerous or more closely intertwined with civilian life. For example, atmospheric, outer-space and underwater nuclear tests produce signals that propagate globally. Thus, the Partial Test Ban Treaty of 1963 could rely completely on national technical means (and did not even mention verification). On the other hand, chemical weapons can be produced not only in the chemical industry, but also in small, covert facilities. Thus, the Chemical Weapons Convention of 1993 contains a detailed verification annex, including a sophisticated scheme for several types of inspections, including challenge inspections. A fictitious ban on certain types of microelectronics or microsystems technology in military systems – say, no microprocessors beyond 16 bits register width and 20 MHz rate, or no inertial-guidance systems below 1 kg mass – would require the right to open and inspect in depth internal details of every military system, an intrusion in military secrecy which would certainly not be acceptable. Also for this reason, limits should rather focus on whole military systems the properties of which can be observed from the outside. Bacteria and toxin molecules are much smaller than microchips, of course, but they would not be used singly, and military applications would rely on laboratories for R&D and on carrier and dispersal systems for deployment and use. This is why the rolling text of the compliance and verification protocol to the Biological Weapons Convention – the negotiations since 1995 have been stopped in 2001 after the USA withdrew from them (Nixdorff *et al.* 2003: Ch. 8)[9] – foresees roughly similar inspection rights as the Chemical Weapons Convention (BWC AHG 2001).

Concerning limits in areas of NT, corresponding inspections are conceivable in case of few centralized facilities. Should cheap production in small units become possible and widespread, then even with anytime-anywhere challenge inspections the required effort may be deemed too high. These arguments would in particular apply to MNT if anybody could 'grow' anything in nearly arbitrary amounts starting with a small device at home. Thus, restricting production based on universal molecular assemblers to few installations would seem to be required. In case of self-replicating nano-systems, where in principle possession of one first unit could lead to a strategic advantage/disadvantage, verification would become extremely difficult. Theoretically, 'nano-monitors' might help if they could recognize one single illegal system, but they would need to be very numerous, and would provide many opportunities for circumvention themselves (see also the considerations on 'active shields' in Sections 1.5.1 and 1.5.5).

Verification of limits on NT-based systems of macroscopic size, down to a centimetre or so, could mainly rely on random inspections of test ranges, production facilities, training activities etc.

The balancing of gains and losses due to limits and the respective verification is a complicated process. Whereas in countries that are strong in military technology the military will rather tend to argue against limits, political decision-makers may be well-advised to follow the precautionary principle, but applied to the international system, avoiding a narrow view of national security through military strength.[10] Consequently, they should ideally rather err on the side of too stringent limits. Due to many counterbalancing factors not only from the military, but also from industry, realistically seen there is no danger of limits that are far too strict.

When considering verification costs, one should keep in mind that present national verification organizations using hundreds of personnel and tens of millions of euros/dollars per year make sure that tens of thousands of soldiers and military equipment worth billions of euros/dollars in potential opponent countries do not exist and thus need not be funded on one's own side as a military compensation. Even more economical is an international verification organization such as the Organization for the Prevention of Chemical Weapons or the Comprehensive Test Ban Treaty Organization (CTBTO), where hundreds of personnel carry out the verification activities on a world-wide scale.[11] Thus, in general adequately verified arms limitations are much cheaper than military preparations for mutual armed conflict.

Of course, it is the decisions of the states to enter limiting agreements or not, and all kinds of political, economic and military arguments will influence the considerations. It is the task of preventive-arms-control research to think through the various options and assess the potential consequences in advance. It is true that the political situation is decisive, thus the outlook for preventive arms control is bleak at present (see Section 7.4). However it is important to remember that political circumstances can change, even if slowly sometimes. Events or public arguments can lead to new insights, and governments do not last forever.

5.3.2 Specificity level

Concerning the degree of specificity of limits, several levels are conceivable. In principle, limits could apply to

- broad or small technology areas (e.g. NT generally, molecular electronics, nanoparticles),
- areas of military application or operation (e.g. close-range surveillance, anti-satellite attack), or
- specific systems (e.g. autonomous combat aircraft, implanted diagnosis and drug-delivery systems).

Previous preventive-arms-control agreements used mainly the second and third options (ABM Treaty 1972–2002; Biological Weapons Convention 1972; Chemical Weapons Convention 1993; Protocol on Laser Blinding Weapons 1995). Limits by technology areas were only used as an exception; one can argue that the nuclear-testing treaties (Partial Test Ban Treaty 1963; Comprehensive Test Ban Treaty 1996) excluded R&D of/with nuclear explosions.[12]

At the level of technology areas there is rarely a civilian–military separation. One consequence is that, should one find that some NT-based areas ought to be limited as such, limits should comprise both sectors. This argument may apply to body manipulation, see Section 5.3.4. In most other technology areas, there are societal or economic interests in using them for civilian purposes, and NT limits should not prohibit such uses. Preventive NT limits should rather work at the level of military application/operation, or the level of specific military systems, not restricting civilian uses at all, or introducing such restrictions that are needed for reliable military limits but do not prevent the important civilian applications.

In many cases, NT will allow new options for fulfilling military missions. Specific limits on using NT could prevent dangers associated with these. In many cases, however, it is rather the military mission as such that produces dangers, even without using NT. In such circumstances, preventive limits should rather focus on the military mission; changes in its carrying out that would come about by NT would be included automatically in a comprehensive agreement. For example, NT-based new biological-warfare agents are already prohibited by the Biological Weapons Convention; NT-enabled small satellites for anti-satellite attack should be forbidden within a comprehensive ban on space weapons, see Section 6.4.6.

Excluding whole military missions or operations will only be possible in some specific cases. Most missions – such as reconnoitring the situation, preventing an invasion (or carrying one out), threatening a state with nuclear attack – could only be removed as part of a fundamental restructuring of the international system. Thus, as long as armed forces exist, new technology for fulfilling such missions cannot be banned by mission, but by specific new systems that would bring new dangers that the international community would rather like to prevent. With respect to NT-based systems, this could apply to micro-robots – independent of purpose or if used as weapon carriers, see Section 6.4.5.

5.3.3 Life-cycle stage

In a simple linear model, new military technologies are generated and introduced in the following stages.[13]

- research,
- development,
- testing,

- production,
- deployment,
- use.[14]

By definition, the goal of preventive limits is to prevent deployment. Thus, one can ban just deployment, for consistency including production and use. An even lesser alternative would be to prohibit only the use of the technology in question. A ban on use was chosen for the Protocol on Blinding Laser Weapons of 1995. Despite its narrow legal scope, the Protocol nevertheless had the effect of practically stopping research, development, testing and acquisition of such weapons. This concerned one very specific technology in which the military interest was limited and which was strongly criticized as violating the rule about unnecessary suffering of the international law of warfare.

With applications where stronger military interest exists, however, the situation will be different. With a ban only on deployment, states could covertly or even overtly do research, development and testing up to a stage where deployment could be decided upon and carried out on very short notice. This could create significant uncertainty for the parties to the treaty, diminishing their confidence in its continuing validity and creating motives to prepare for a breakout. In general, preventive limitations that ban only deployment suffer from an in-built undermining tendency.

To strengthen a treaty and raise the confidence in its stability, one would thus prefer limits that apply as generally and as early as possible. Not only should production be included, but wherever feasible also testing and development. For comprehensiveness one could argue for limits starting already at the research stage; however, here the applications are usually open, and limits on research would constrain technologies with potential benefits to society. In addition, there are strong economic and other motives for developing goods for the civilian market; thus, strong economic and political resistance is to be expected if limits become too general. This holds for most technologies, and in particular for NT with its very broad scope.

Due to the openness of research, limits on it would pose difficult problems of definition and verification. Thus generally, NT limits should apply to development and the later stages. Even though in this case applied research geared towards banned military applications would remain legally allowed, one can assume that, with well-defined limits on development, testing and deployment, together with adequate verification, such research will nevertheless be curtailed as a consequence of funding decisions taken on the respective national level: if something cannot be developed or deployed, then spending much money for its research is a waste. Even if research is not reduced, development and testing usually take a significant time and could be detected by adequate verification, so that a fast breakout can be excluded. Of course, transparency measures on research would increase the confidence in the treaty.

In very specific instances, limits on research are recommended, however. In NT this would apply, e.g. in order to prevent self-replicating systems from being created in the laboratory that might later be impossible to contain. The same would hold if a detailed study identified a clear, strong threat from very small nuclear weapons.

In case of agreed limits on research, verification presents difficult problems, and the methods and procedures will have to be all the more intrusive, the smaller the research objects or tools are. Similarly to genetics research, verification of limits on NT research will require access to, and co-operation of, the laboratories – but the same holds already for development and testing activities.

As the debate on the verification Protocol for the Biological Weapons Convention (that bans development etc., but not research) shows, industry and the government in the USA have developed strong objections to on-site inspections in laboratories,[15] even though other countries leading in pharmaceutical and genetics research and technology have no such reservations.

One should keep in mind that a political assessment needs to take into account the potential outcome. Rejecting limits due to the intrusiveness of verification will avoid industrial espionage by inspectors. However, the ban on biological warfare could break down if the Convention remains unverified in the face of fast advances – accelerated by NT – in molecular biology. In the end this could result in much higher threats than the risks following from strict verification.

5.3.4 Form of civilian involvement

In principle, there are two ways of dealing with civilian R&D in the context of preventive arms control. The first is to leave the civilian sector unregulated; it is only advisable where there is practically no civilian interest or application, as at present with nuclear explosions (after the cessation of so-called peaceful nuclear explosions). Usually, however, civilian and military technologies and R&D are intertwined in many ways so that the civilian sector should be included. In this second case, the civilian limits should normally be specific, different from the military ones. Only rarely can the limits be identical, as proposed for non-medical implants in Section 6.4.3.

Most *research* is inherently ambiguous and the new knowledge gained could be used for civilian as well as military purposes; *development* of actual systems is usually specific of the respective sector. However, there are areas of science and technology where the same equipment or laboratories can even be used for development in both sectors. Chemical facilities for R&D of insecticides could also be used to work on chemical warfare agents. In some cases, civilian biochemical/medical research (and therapy) uses the very same toxins that are outlawed for non-peaceful uses (e.g. Rossetto *et al.* 2004). To investigate the pathogenic and virulence mechanisms of microorganisms, medical research purposefully modifies their genes which can create new agents that could be used for biological warfare (Nixdorff *et al.* 2003: Ch. 11). In aviation, much of military as well as civilian R&D takes place in the same institutions and firms, with knowledge flowing in both directions (Altmann 2000; see also Brzoska 2000 and refs).

These civilian-military ambiguities and links are the reason why the Chemical Weapons Convention includes verification in civilian chemical industry, and why the rolling text of the compliance and verification protocol to the Biological Weapons Convention – now blocked – foresees visits in biological facilities (BWC AHG 2001).

Because civilian R&D could be used for covert military work, limits on military R&D will usually have to be complemented by verification in civilian R&D institutions.

In order to prevent circumvention of limits on military R&D, in many areas one should furthermore devise limits for civilian R&D as well. Specific civilian limits will in most cases be wider than the military limits. One motive is beneficial civilian uses, another general scientific curiosity that wants to go beyond known parameter ranges. Misuse of civilian R&D for military purposes will have to be prevented by strong verification rights – which should be easier to accept for civilian activities. With respect to NT, this might apply to small satellites; a significant number could be allowed for civil Earth monitoring or space research if subject to intensive licensing and inspection procedures while military small satellites would be strictly limited.

Wherever possible, one should modify the civilian technologies so that their benefits accrue while the possibility for military misuse is minimized. One example for such peace- and stability-oriented shaping of technologies is the conversion of research reactors from highly enriched uranium to low enriched material that cannot be used for a nuclear bomb (proliferation resistance) (see e.g. Glaser 2002 and refs). Another would apply to large satellites for converting solar energy; if microwaves instead of a laser of much shorter wavelength were used to transmit the energy down to Earth, the laws of wave propagation (namely diffraction) would prevent a narrow focus that could be used for attack.[16] Concerning MST/NT, where one might want a ban of mobile systems below 0.2–0.5 m, for important civilian exceptions, e.g. surgical micro-robots, one should demand a design with limited mobility, remote control and external inductive power supply (see Section 6.4.5 and Altmann 2001: Sections 7.2, 8.2).

Civil society will have its own reasons to limit new civilian technology, to protect humans, the environment or societal processes. Such national rules and the checks on observance may serve as a precedent for the internationally agreed limits and their verification, and both should be co-ordinated with each other.

There is also an indirect effect on the military of civilian R&D and widespread use of a technology: improvements can reduce its cost markedly ('learning curve'), allowing much wider military uses or new military applications. The prime example is digital computers – one of the areas where the military are now routinely using commercial-off-the-shelf products. Because this effect concerns civilian mass products, it would be very difficult to contain by preventive arms control.

6

PREVENTIVE ARMS CONTROL CONSIDERATIONS FOR NANOTECHNOLOGY

In this chapter, a first assessment of nanotechnology (NT) within the framework of preventive arms control is done. These criteria are applied to the potential military NT applications in Section 6.1, and to those that are conceivable with molecular NT – in general terms – in Section 6.2. The summary evaluation in Section 6.3 arrives at seven potential NT applications where preventive limits are highly important. Options for the design of such limits are discussed in Section 6.4, considering also positive uses and verification aspects. The final Section 6.5 presents a few meta-aspects of preventive arms control that have come up in the present work.

6.1 Applying the criteria to NT

This section discusses the potential military NT applications, presented in Chapter 4, under the criteria set of Section 5.2. In some cases, fulfilment of a criterion is not NT-specific (e.g. new biological warfare agents could come about by 'traditional' molecular-biology research and development (R&D)); however, NT will significantly accelerate R&D in many areas and may make many applications much more economical.

The systematic run-through here follows the sequence of the criteria (different from the order by application used in my microsystems-technology (MST) analysis (Altmann 2001: Ch. 6)). It serves to find out the problematic applications. For a simple overview presentation, a table will be filled in Section 6.3, using four symbols to designate danger (−), no significant effect (0), improvement (+), or an unclear estimate (u). Even with such a simple, three-step classification scheme, a certain arbitrariness in assigning these values to particular technologies or military applications is unavoidable, e.g. due to special cases or exceptions.

In many cases one could make arguments as to why a technology or application might indirectly influence a criterion area, maybe only in special circumstances. Because this is complex and often ambiguous (e.g. improved data-processing capabilities could increase or decrease stability, depending on the circumstances – better information on an opponent could promote one's own surprise attack, but also dispel fears about the other's surprise attack), and because the context is whether preventive-arms-control restrictions should be introduced, the considerations focus on the direct and more or less obvious consequences – indirect consequences are only considered with the criteria group III on humans, environment and society that deal mostly with peacetime consequences (the other criteria are about prevention of war, a few about conduct of war). (Of course, the dividing line between direct and indirect effects is fuzzy.)

In a few areas the evaluation differs according to specific applications. Thus, sub-groups are used as follows:

- Sensors: for general purposes, for the battlefield and for verification;
- New conventional weapons: metal-less small arms and light weapons, small guidance and navigation systems, armour-piercing projectiles and shaped charges, and small missiles;
- Autonomous systems: unarmed and armed ones – the differentiation hinges on the possibility to discriminate between the two types. In case unarmed systems could be converted to armed ones easily, then the assessment for the latter applies.
- Nuclear weapons: auxiliary systems, computer modelling and very small nuclear weapons.

6.1.1 Criteria group I: adherence to and further development of effective arms control, disarmament and international law

6.1.1.1 Prevent dangers to existing or intended arms-control and disarmament treaties

The arms-limitation criterion is not directly affected by the more generic applications (from electronics to propellants/explosives) and by several military-specific ones: camouflage, distributed sensors for general purposes and on the battlefield, armour/protection, soldier systems, unarmed autonomous systems and auxiliary systems for nuclear weapons, chemical/biological protection.

Distributed sensors for treaty verification could have positive effects in arms-limitation treaties. This could concern better detection of chemical or biological warfare agents or of their effects in the body, e.g. of production personnel. Also, smaller/cheaper/more numerous sensors could help in monitoring satellites in outer space, space launchers, missiles, aircraft and military land vehicles on earth. They could also contribute to verification of limits on small systems. Of course, separation from battlefield sensors is not always easy; stationary deployment may help.

Negative effects are possible with new conventional weapons if they are used to circumvent the Treaty on Conventional Armed Forces in Europe

(CFE) or future regional agreements for the control of small arms and light weapons. If new weapons contain less/no metal, verification of such agreements (by induction detectors or x-raying machines) would become more difficult. The same argument would hold for small missiles. The CFE Treaty could also be undermined by armed autonomous systems. Armed mini-/micro-robots could have the same effect and could also endanger the Anti-Personnel Mine Convention, since due to their mobility and other characteristics they would not qualify as a mine, but could nevertheless function as such.[1]

Small and/or more autonomous satellites, if used for anti-satellite attack, would counteract the general ban on space weapons that the international community has striven for since decades.

Vastly extended computer modelling of nuclear weapons can lead to new warhead designs that could on the one hand undermine the Comprehensive Test Ban Treaty. If the models became so detailed – and maybe verified with data from historical tests – that confidence is sufficiently high, then the new designs might even be built and deployed without tests. While formally complying with the Comprehensive Test Ban Treaty, this would violate at least the spirit of the Nuclear Non-Proliferation Treaty that stipulates good-faith negotiations towards nuclear disarmament. The same holds for very small nuclear weapons. New chemical or biological weapons would contravene the Chemical and the Biological Weapons Conventions.

6.1.1.2 Observe existing norms of humanitarian law

The criterion about the law of warfare also is not directly affected by the more generic applications and by several military-specific ones: camouflage, all types of distributed sensors, armour/protection, soldier systems, unarmed autonomous systems, mini-/micro-robots without weapon function, systems for outer space, auxiliary systems for and computer modelling of nuclear weapons, chemical/biological protection.

New types of conventional weapons could in principle contravene the rule against superfluous injury or unnecessary suffering, e.g. if NT-enhanced projectiles act similarly to dum-dum bullets. Since states are under an obligation to check consistency with humanitarian law for all new weapons, means and methods of warfare (Art. 36 of Additional Protocol I to the Geneva Conventions of 1949 (ICRC 1977)), one can expect caution here. Nevertheless, the international community should monitor this area and take action, if required, in particular because there may be differences of opinion.[2] On the other hand, NT will allow much more precise targeting with correspondingly reduced destructive power, so that wounds could be smaller and collateral damage less, with a positive effect on humanitarian law.

Body manipulation and implanted systems could have negative results if aggressiveness in fighting were increased by electrical or hormonal stimulation to such an extent that protection of non-combatants or combatants *hors de combat* is no longer warranted.

Great problems can be foreseen with armed autonomous systems, including mini-/micro-robots that are armed or are used to designate targets. Discriminating between combatants and non-combatants, recognizing legitimate military targets or recognizing the condition *hors de combat* (wanting to surrender, unconscious or unable to fight by injury or illness) are complicated tasks that require a more or less human level of intelligence. For quite some time – at least one decade – artificial systems will not reach such a level. At least until then, autonomous decisions about weapon use and autonomous picking of targets for attack by other weapons would contravene humanitarian law. The scale would increase with the number of armed autonomous systems and mini-/micro-robots. This problem can be avoided by demanding a human in the decision loop at least for the decisions whom/what to target and when to attack, i.e. remote control, restricting autonomy to the other phases of the mission. However, this would only work if technological advance is limited. It is easy to foresee circumstances where military necessity would dictate fast decisions that would overstrain a remote human controller. This would particularly hold true if the opponents' systems can act fast and are numerous. If one's own systems multiply, there will also be strong economic motives not to tie valuable human time to every single weapon-activation decision. Finally, with mostly autonomous systems one also has to plan for the contingency that the remote-control communication link may be jammed or broken. Humanitarian law would require a block on weapon use in such a case if there is a danger of attacking illegitimate targets. Upholding such a rule only seems realistic if the numbers of weapons/carriers stay at about current levels, not with swarms of fighting mini-/micro-robots.

Hypothetical very small nuclear weapons, even though much below the explosive yield and radioactive fallout of the traditional large ones, could still create unnecessary suffering and affect non-combatants disproportionately. New chemical or biological weapons would violate the rules banning such methods of warfare.

6.1.1.3 No utility for weapons of mass destruction

Whereas several generic and more specific NT applications could be used for weapons of mass destruction – e.g. vehicles or robots as new carriers – direct utility exists only in a few areas. New auxiliary systems would not change the basic properties of nuclear weapons. Some increase in yield-to-mass ratio and more sophisticated targeting procedures or triggering mechanisms are conceivable. On the other hand, safety and security might be improved somewhat over the existing permissive action links.

As mentioned in Section 4.1.19.2, much more powerful computers will allow markedly improved modelling of nuclear explosions that could lead to new warhead designs. The hypothetical very small nuclear weapons, while blurring the distinction between conventional weapons and weapons of mass destruction, would at least in the upper end of their yield (above, say, 10 t TNT equivalent) nevertheless still qualify as the latter.

New chemical and biological weapons would count as weapons of mass destruction if they can attack many target humans, animals or plants. Even if NT and biomedical advances could provide mechanisms for very selective uses, maybe even against only one individual, the knowledge gained in the R&D could be used for agents and carriers targeting larger collectives.

Chemical weapons that would confuse large numbers of people for a sufficiently long time could indirectly, by the breakdown of societal production and distribution systems, cause mass death.

Positive uses helping in the disarmament of weapons of mass destruction or reducing their effects can ensue from sensors for treaty verification and from material for protection from and neutralization of chemical or biological agents.

6.1.2 Criteria group II: maintain and improve stability

6.1.2.1 Prevent destabilization of the military situation

In this context, stability refers to the situation between two (or more) actors that see themselves as potential opponents in armed conflict so that military preparations are part of the mutual relations (even if that is not spelled out clearly all the time). General military stability exists if neither side can expect a successful outcome if it started war. If the mutual relations deteriorate, the short-term notion of crisis stability becomes important. It refers to the possibility of mounting a surprise attack, in particular reducing the opponent's armed forces so that success in defence becomes questionable; in nuclear strategy, the question is whether a second strike can still be delivered (deterrence). When one side fears a successful surprise attack, it is under pressure to pre-empt that by itself attacking first. A situation where both (or more) sides have strong motives for pre-emptive attack is highly unstable.[3] Special cases concern stability in armed conflict with respect to escalation to qualitatively higher levels, in particular to employment of weapons of mass destruction.

Direct effects on military stability are not seen for many of the more generic applications. However, much more powerful computers could be used for more precise targeting which could improve the outlook for surprise attacks. Much increased capabilities in artificial intelligence, used in battle management, could lead to a higher reliance on automatic decisions whereby opponents in a crisis might faster slide into war; however, in particular at the pre-conflict stage, humans will try to keep the situation tightly under control. Faster propulsion and lighter, more agile vehicles can reduce the travel and thus warning times.

Among the more specific applications, camouflage, general distributed sensors, armour/protection, soldier systems and unarmed autonomous systems have no great direct effects. Battlefield sensors would help to locate targets, making easier fast precision attacks. Sensors for verification, on the other hand, would serve in the limitation of weapons and capabilities for offence, contributing positively to stability. New conventional weapons with higher speed and more precise targeting would have some negative effect. Strong destabilization would result if conventional attack against nuclear weapons carriers and strategic command-and-control systems became available (see Miasnikov 2000). Body manipulation and implants that reduced reaction times and linked brains with computers could have some destabilizing effect, but the other factors and the overall setting would be much more important.[4]

Armed autonomous systems would act and react with very short time lines. Deployed at close mutual range, strong pressures for fast attack would be at work between opponents of roughly comparable technological capabilities. Movements or actions mistaken as (pending) attack could easily lead to the start of hostilities. The same holds for accidents or errors in the warning systems. Even stronger destabilization would follow from mini- and in particular micro-robots that are armed or are used to designate targets, because they could be sent covertly already before armed conflict, ready to strike (or guide strikes) at important nodes from within at any time. Surprise attack on the command, control and communication system, e.g. on sensors, communications, weapons-control equipment, could drastically reduce one's capabilities. The warning time could be extremely short. In parallel, strong motives would exist to send mini- and micro-robots for reconnaissance into the opponent's territory, some in advance, others at the time. Even if some of them were noticed, differentiating between systems for reconnaissance and for direct attack would be extremely difficult. As a consequence, armed autonomous systems and mini-/micro-robots at large would create great uncertainty; human commanders would be correspondingly nervous, and equipment would be programmed to operate fast. Unintended action-reaction cycles between the mutual systems of warning and attack have to be feared.

In outer space, (swarms of) small satellites could provide improved detection capabilities for targets on the ground and in the air, and more capable communications channels, that could be used for surprise attack on Earth. Much more destabilizing would be their capacity to attack other satellites, either by collision at high relative velocity, or by manipulation after rendezvous and docking. This could knock out the various reconnaissance, early-warning and communication satellites that are of central importance for warfare on Earth for the largest military power(s), including command and control of nuclear weapons. Also, the communication satellites that have become central for civilian life could be destroyed. The potential for surprise attacks in space would increase if small satellites were to be deployed in high numbers; particular risks would exist if they were sent to the altitude of geostationary orbit (35,800 km). High-impact destruction needs tests that would be observable from the ensuing debris and trajectory change. Tests of docking and subsequent manipulation could, however, be done under the guise of maintenance and repair. Small satellites for docking could 'shadow' the target satellite, ready to close in for attack at any time. A traditional satellite may not even be equipped with sensors to detect a close neighbour. In the future, such sensors would be used; defence preparations could include mechanisms for preventive attack of anything approaching within, say, 1 km. Whereas the strongest destabilization is expected if small satellites are already deployed in space, they could also be lifted there on short notice by small launchers. These would not depend on large space centres, but could take off from nearly anywhere, including ships and aircraft. Their travel time to low Earth orbit (300 to 2,000 km) would be many minutes, to geostationary altitude a few hours, with corresponding warning times.

With respect to nuclear weapons, no great effects are expected from new auxiliary systems or computer modelling. However, direct reduction in stability, in particular regarding deterrence against nuclear-weapons use, would follow from very small nuclear warheads. With a blurred distinction from conventional weapons, the reluctance to use them would be lower than with traditional nuclear weapons, making actual employment more probable. This crossing of the nuclear threshold could act as a precursor to the use of larger nuclear weapons.

In a similar vein, new chemical or biological weapons that would seem to make their use more manageable, targeting only the intended group(s) and saving one's own forces and population, could undermine stability.

Material for protection from and neutralization of chemical or biological agents can have different consequences for stability: expecting a well-protected opponent, one side might prefer to forego chemical and biological warfare. If one feels well protected oneself, on the other hand, one would be more inclined to use chemical or biological weapons. That is, the judgement depends on the context, in particular on the existence and effectiveness of selective agents and their availability (or the perceptions thereof) for the respective parties.

6.1.2.2 Prevent technological arms race

If one among several potential opponents introduces new technologies into its armed forces, strong motives to do the same exist for the others. This holds in general and – because of the long lead times – works already at the stages of R&D. As discussed in Section 5.1.4, a similar mechanism is in effect even between partners in a military alliance.

With respect to the various potential military NT applications, arms-race motives exist practically for all of the more generic and all of the more specific uses. Nearly everywhere there is the promise that traditional military missions can be fulfilled better or that new ones become accessible. Arms-race motives are particularly urgent where threats to own forces or restrictions of their operations can be foreseen. If applications have a more defensive character or are only used in auxiliary functions, there may still be motives to introduce them, too, but the need to overcome them could be less. The former could apply to soldier systems that provide medical functions and to chemical/biological protection and neutralization. The latter could hold for sensors for monitoring and maintenance of equipment, for logistics and environmental and security surveillance around fixed sites, and for auxiliary systems in nuclear weapons. There is only one category where an application would act positively, against arms racing: sensors for verification of arms-limitation treaties.

6.1.2.3 Prevent horizontal or vertical proliferation/diffusion of military-relevant technologies, substances or knowledge

Non-proliferation became a goal of arms control with the rise of nuclear weapons. Here, horizontal proliferation means that additional countries acquire/build nuclear weapons whereas vertical proliferation refers to states with nuclear weapons acquiring more or new types of them. When translating non-proliferation of nuclear weapons into more concrete terms, one deals with ready-made systems (actual bombs or their components), technologies (such as uranium reactors for plutonium production, ultracentrifuges for uranium enrichment), substances (in particular, plutonium and highly enriched uranium) and knowledge (e.g. about explosive lenses for spherical-symmetric implosion of plutonium pits). Diffusion refers to a less tangible process where technologies, substances or knowledge spread – by civilian uses, education of foreign students, science exchange etc. These concepts can be generalized to other military-relevant technology areas. Missile technology is an obvious example that is connected to weapons of mass destruction. In a more general vein, one can point to computing technology.[5]

Concerning potential military uses of NT, vertical as well as horizontal proliferation of systems, technologies, substances and knowledge has to be feared in all areas. Transfer of systems will be easier the smaller they are. Diffusion will take place particularly in the more generic fields where widespread civilian applications are foreseen. Similarly to the arms-race criterion (Section 6.1.2.2), there are very few uses where proliferation motives exist, but might not be judged negatively if they have auxiliary or rather defensive functions: soldier systems with medical functions, distributed sensors for monitoring and maintenance of equipment, for logistics and environmental and security surveillance around fixed sites, auxiliary systems in nuclear weapons and chemical/biological protection and neutralization. Also here there is only one area that could counteract proliferation: sensors for verification of compliance with treaty limits.

6.1.3 Criteria group III: protect humans, environment and society

6.1.3.1 Prevent dangers to humans

This criterion applies mainly to peacetime – it is about (new) military technologies or substances negatively affecting people, mostly in civilian society. Armed conflict is by definition connected to dangers to soldiers and practically always brings dangers also for non-combatants – but these aspects are being dealt with by the criterion about the law of warfare, see Section 6.1.1.2.[6]

Direct dangers to humans are not foreseen for most of the military NT applications. New materials could bring risks to human health, e.g. by nanoparticles travelling across boundaries in the body and interacting with cell processes; the same could apply to new propellants and explosives. For each such material research is needed to elucidate potential harmful effects. New conventional weapons could bring dangers to humans if they proliferate to criminals; this could apply to small arms and maybe light weapons. In particular, metal-less arms or small missiles would provide new options for terrorist attacks.

Armed autonomous systems would probably be reserved for armed conflict – internal law-enforcement institutions would likely keep a human in the decision loop all the time for constitutional reasons. However, the constitutional argument does not hold in operations other than war, e.g. in an occupied country, so there is a grey area here where humans might become unjustifiably endangered. Armed and unarmed autonomous systems could also cause accidents in peacetime: testing would probably be limited to military property or, in international territory, to specific areas with advance warning. However, there would also be routine patrolling operations along borders, on and in the oceans and in the air. Collisions with other vehicles or crashes could result in loss of life.

Mini-/micro-robots acting as weapons would produce dangers if they became available for criminals, in particular terrorists.[7] The same would apply to new chemical or biological weapons. Attacks could be directed at single persons – small robots could sneak into a building, injecting a drug during sleep, firing a small projectile or setting off a small explosion close to a person's body. Sophisticated chemical or biological agents could be used against selective targets or for mass attacks, e.g. poisoning drinking water or food. If new weapons and autonomous armed systems could be strictly limited to the military, they would not count under this criterion. With chemical and biological weapons, however, there is the additional risk of unintended release from military production or storage facilities.

Very small nuclear weapons – if feasible at all – would probably be guarded as strictly as larger ones so that use by criminals and terrorists in peacetime would not create qualitatively new concerns.[8]

Body manipulation and implanted systems could produce dangers to soldiers in peacetime if they bring about secondary effects. More important could be the indirect consequences if military R&D and application of body manipulation contributed to general use in society with too few restrictions (see Section 6.1.3.3). Many negative secondary effects on body and mind are conceivable, similar to and beyond those of addictive drugs.

A potentially positive effect would exist if sensors developed for treaty verification also monitored dangerous substances and were transferred to civilian society. Systems for protection against and neutralization of chemical or biological warfare agents, used in cases of terrorist attacks, would reduce dangers to humans and thus count positively. However, sensors and protection/neutralization systems would probably be more useful in a civilian context, if they were being developed in a civilian setting, not primarily spun off from military R&D.

6.1.3.2 Prevent dangers to environment and sustainable development

This criterion also deals mainly with peacetime effects, since in armed conflict damage to the environment can hardly be avoided. However, it is important to note that on the one hand, the Environmental Modification Convention of 1977 bans military and hostile use of environmental-modification techniques having widespread, long-lasting or severe effects (ENMOD 1977). On the other hand, the Additional Protocol I to the Geneva Convention stipulates in Art. 35 that in armed conflict '[i]t is prohibited to employ methods or means of warfare which are intended, or may be expected, to cause widespread, long-term and severe damage to the natural environment' (ICRC 1977). Thus, here aspects of warfare do enter the deliberations.

Similarly to the preceding criterion, most military NT applications would bring no direct dangers to the environment. New materials as well as propellants and explosives could damage life processes in plants and animals; clarification will need research on each substance. More efficient energy sources and propulsion engines as well as lighter vehicles could save energy and resources consumed by the military and would thus rather contribute to sustainability – but only if the reduction is not compensated for by an increased number of military vehicles, robots etc.

Very small nuclear weapons would still produce and release radioactive material that would pollute the affected area. New, highly selective chemical and biological weapons that would be used only in minute quantities could leave the environment untouched; however, others could have negative effects, e.g. if directed against widespread plants.

Positive effects are to be expected if distributed sensors for treaty verification also monitor substances that are dangerous to the environment. In so far as chemical or biological weapons pose a risk to the environment, protection and neutralization systems can count as positive.

6.1.3.3 Prevent dangers to the development of societal and political systems

This criterion deals mainly with the internal processes within societies in peacetime. These will mostly be affected by military technology applications in an indirect way. This criterion goes beyond the traditional ones that focus on the direct effects; it was added by the German joint projects on preventive arms control when they discussed more general technologies such as MST (Altmann 2001: Ch. 6; Neuneck and Mölling 2001).

Many of the more generic military NT applications would not affect society. Even vastly more capable computers and communication devices in the military, together with corresponding software, need not have an effect. However, if these technologies were used for tracking and surveillance of individuals, which would become possible in the context of ubiquitous computing, they would infringe on freedom and privacy.[9] Much more capable artificial intelligence could be used to mine databases for information about individuals, corporations and other institutions.

Among the more specific military applications, no significant dangers would follow from camouflage, armour/protection, soldier systems, armed and unarmed autonomous systems, small satellites and launchers, auxiliary systems for and computer modelling of nuclear weapons. This finding assumes that the corresponding systems and weapons remain under strict state control.

Distributed sensors for general purposes (such as monitoring and maintenance of equipment, logistics, surveillance around fixed sites) provide some capability of tracking and surveillance. Such (mis)use would be much easier with sensors developed originally for battlefield monitoring, suitable to be scattered, some extremely small, competent in finding and recognizing targets at a distance. They could be used for eavesdropping and tracking of individuals and vehicles at least by state agencies. Should the sensors become cheap and numerous, also private entities and criminals could use them, for private espionage or for other crimes. With immobile sensors, the capabilities would be limited for non-state actors, however. Mobile mini-/micro-robots would provide a markedly higher potential for intrusion into privacy and eavesdropping. If small enough, they could covertly enter offices or houses – even through the crack under the door, or through an open window. Small robots could also act – e.g. steal, disrupt or destroy something. Injuring or killing people could be done while they sleep or, similar to an insect – and maybe even using an insect(-like) body – at any time and any place, in public or private. Tracing back to the originator could be very difficult. If such possibilities arise, one can expect countermeasures such as seals for openings and detectors for small mobile objects. Nevertheless, a feeling of endangered privacy and insecurity is probable with a high burden on society. Negative influences on democratic processes can be expected, too.

If new conventional weapons fall into private hands, security in society could suffer. Checking for arms that are not visible by x-rays and metal detectors would require more intrusive procedures.

Body manipulation and implants in the military would at first affect only a small part of society. Soldiers might accept them voluntarily because they would provide better fighting efficiency or faster treatment after injury. After some time, or in countries with less respect for human rights, soldiers might be ordered to get specific drugs or implants, similar to preventive vaccinations.[10] However, this touches fundamental questions on whether non-medical body manipulation should be allowed, and if so, what types. Up to now, medical as well as general ethics treats the human body as a sanctuary; invasion is only justified if there are medical reasons, that is for therapy – trying to restore a state of health. Enhancement is not yet really possible which contributes to the reluctance towards it. Concerning genetic improvement – that would be possible already today – there is general condemnation. In sports, doping is penalized. However, the rejection of body improvement is not absolute. Voluntarily, some people have cosmetic surgery to improve their looks or adapt it better to the dominant aesthetics in their society. Some males accept risks of secondary effects in order to prevent age-related weakening of their sexual performance. Some parents accept chemical treatment of their hyperactive children. Use of drugs that enhance alertness and learning among students is more widespread than one would think (see Hall S.S. 2003).

Many kinds of implants and other body manipulation are conceivable in the future. Several will be developed to treat diseases or injuries. The understanding of organ function will grow as will the experiences with long-term implants. Thus, the potential for enhancement of healthy people will grow. Some applications might be considered useful, e.g. an implanted chip with medical data; this is already possible today, radio-frequency identification chips are implanted routinely in pets in several countries. In one to two decades, a general-purpose processing and communication device for wireless access to the general network might be possible that contacts the brain via a large number of microelectrodes. If that were to come, memory enhancement should become feasible as well. It is equally easy, however, to conceive of many possibilities of abuse and severe dangers. Wireless identification devices could be used to track people. Corporations providing the 'program' for implanted sensory and emotion devices could manipulate people. Electrical stimulation of a reward or (sexual) happiness centre in the brain, or release of endorphins, could make people virtually addicted, reducing their actions to those needed for bare survival, as with the rats in the famous experiments of the 1950s (Olds 1958). Enhanced organs, delayed ageing etc., even if individually attractive at first sight to many, need not be judged positively if the overall societal outcome is taken into view. The changes in human nature that may become possible are profound. Ethical questions of similar type as with stem-cell research or human cloning will be posed, but on a wider scale and with higher importance.

All these are purely civilian problems. The extent to which non-medical body manipulation should be allowed should be considered and debated in society at large and its democratic decision-making bodies. However, widespread military use could undermine and pre-empt an open-ended societal debate by accustoming people to the idea and its realization. Military body enhancement would create facts that may be difficult if not impossible to revoke. Thus, it poses dangers to societal and political systems.

Should very small nuclear weapons become possible, great dangers to societal and political systems could ensue, but only in case they became available to non-state actors. While this possibility is doubly hypothetical, new chemical and biological weapons would offer many possibilities for terrorist attacks and extortion with the corresponding dangers for society. Chemical and biological protection and neutralization, on the other hand, would reduce dangers arising from old or new agents, leading to a positive assessment.

6.1.3.4 Prevent dangers to the societal infrastructure

As the other criteria in this group, this one mainly concerns peacetime. The infrastructure provides basic services, such as transport and communication, and goods, such as water and electricity. Massive disturbances can ensue from breakdowns. Thus, the societal infrastructure may be a target in particular for terrorists.

For the more generic military NT applications, no direct effects on the societal infrastructure are evident. Of the more specific uses, dangers do not follow from camouflage, battlefield sensors, soldier systems, body manipulation, unarmed autonomous systems and mini-/micro-robots without weapon function. General sensors and those for treaty verification, if employed at infrastructure installations, could help in the protection against intruders and in the detection of manipulations and of substances in drinking water. Similarly, military armour/protection technologies could be helpful in protecting against attack by terrorists.

Among new conventional weapons, metal-less arms, small guidance and small missiles, if diffusing into the hands of criminals, could be used for more effective/more secretive attacks on the societal infrastructure. The same holds for mini-/micro-robots – those with weapon function could be used directly, those without weapon function could nevertheless be used to transport highly toxic agents or damage central components after entering into installations. Small satellites and launchers could attack commercial satellites that fulfil an important role in the global communication infrastructure – such attack could also affect states that are not a party to an armed conflict. Armed autonomous systems would not easily become available to criminals. However, they could damage infrastructure by accidents, in particular by crashes from the air.

Very small nuclear weapons, if feasible, would present strong dangers to infrastructure, but only if they became available to non-state groups. New chemical and biological weapons could be delivered via drinking water and could thus be used to shut down the supply. Chemical/biological protection and neutralization would act in the opposite direction.

6.2 Preventive-arms-control criteria and molecular NT

Because molecular NT (MNT) is not imminent and its military uses can only be described in general, speculative terms, the consideration of MNT under the criteria of preventive arms control is general and cursory. Several aspects have been discussed by Gubrud (1997).

6.2.1 Criteria group I: adherence to and further development of effective arms control, disarmament and international law

MNT production of nearly unlimited numbers of armaments at little cost would contradict the very idea of quantitative arms control. Practically all types of new MNT-enabled weapons would present *dangers to existing or intended arms-control and disarmament treaties*. This holds for limitations on conventional (heavy, small, light) arms and land mines, and a space-weapons ban. Artificial intelligence and cyborgs, not covered by treaties up to now, might nevertheless create problems in particular concerning limits on personnel. Particular dangers for the Chemical and Biological Weapons Conventions would ensue from self-replicating nano-robots with selective action in the body. With its potential for pre-emptive attacks on nuclear weapons and simultaneously for massive scale-up of their production, MNT would also undermine nuclear-weapon limitation treaties, not to speak of nuclear disarmament. Replicating nano-robots could be used to create widespread, long-lasting and severe changes in the environment, which would violate the Environmental Modification Convention; general conversion of the ecosphere to grey goo would of course go way beyond any intended military effect. MNT would provide chances for better treaty verification, e.g. by ubiquitous nano-robots that monitor practically everything. Of course, this would be highly ambiguous and would bring dangers to humans and society by itself.

Concerning *existing norms of humanitarian law*, some of the concerns mentioned in Section 6.1.1.2 for 'plain' NT could be reduced by MNT. Because actual fighting would largely be carried out by artificial systems, manipulation of soldiers' bodies leading to over-aggressiveness would be less of a problem. Assuming (super-)human artificial intelligence in armed autonomous systems including mini-/micro-robots, non-combatants or combatants *hors de combat* could be recognized reliably. This could be different in small, maybe self-replicating entities. Humanitarian law could be endangered on an unprecedented scale, if artificial intelligence used in warfare did not obey these rules – and this could come about by its own 'conscious' decision, deviating from the goals of the original 'programmers'. With possibilities to manipulate body and brain by invading nano-robots, many ways of transgressing the legitimate goal of rendering combatants unable to fight are conceivable.

In many respects MNT could be used for *weapons of mass destruction*. Production of traditional chemical, biological and nuclear weapons could be drastically increased. New MNT-based chemical or biological weapons could act selectively, but massive destruction could also be built in – or could evolve against the original intent after release during generations of self-replication. Grey goo, if used as a weapon, would lead to extinction of all life forms, even if humans or animals were spared from direct consumption. Attacks on people's brains that would 'only' confuse them could lead to breakdown of society and massive loss of life. Positive effects from better sensing of and protection against weapons of mass destruction that MNT may also bring are clearly of much less importance.

6.2.2 Criteria group II: maintain and improve stability

MNT would strongly *destabilize the military situation* between potential opponents (Gubrud 1997). In international territory such as the oceans, surveillance and fighting systems of all sizes would be interspersed with each other; in outer space, motion is restricted by the orbital laws. Attacks could start here with zero warning time, and systems would probably be programmed for fast action in case of unclear events. For transport across land borders, longer warning times would apply. However, covert pre-deployment in central installations with the potential to strike at any time would have to be feared, and would be used as a counter. In case of systems that self-replicate in the wild, extremely small causes (one first copy) could have extremely strong consequences. Thus, extensive reaction is plausible on detection of one such cause, and action-reaction cycles in case of erroneous detection could lead to the most serious consequences. With respect to nuclear stability, nuclear-weapon states would have to fear disarming attacks by nano-robots penetrating the warheads and carriers, together with efforts at reducing the effects of a response strike by missile defence and civil defence. This would strengthen motives for early use and would increase the chances for unintended nuclear war.

With its potential for extremely fast increases in military production, MNT would represent the culmination in *quantitative technological arms races* (Gubrud 1997). In *qualitative* terms, extremely fast change could ensue from autonomous R&D of new military systems by (super-)human artificial intelligence. Surprises would have to be expected at any time. Early start could provide decisive advantages, up to – in theory – the capability of world domination (Gubrud (1997) has written of a 'new type of strategic instability: early advantage instability'). As a consequence, the military powers would be under extreme pressures to go forward as rapidly as possible when MNT drew close, and thereafter. Because of the potentially huge consequences, even political/military partners might not want to rely on the stability of their mutual relations. Mistrust about what a partner could do with some lead is probable, so technological arms races would not only develop between potential opponents.

The huge military potential of MNT would also create strong motives for *horizontal or vertical proliferation/diffusion of military-relevant technologies, substances or knowledge*. Export of just one copy of a self-replicating weapon nano-robot would suffice to create billions of them. Transfer of one universal molecular assembler and the appropriate synthesis programs would provide the hardware and software for arbitrary armaments growth. If assembly could take place at home or in small businesses using everyday ingredients, weapons, including nano-robots for offensive action in the body, could become available to individuals, not just states. 'Molecular hackers' could produce the real-life equivalent of computer viruses. Whereas the latter directly affect only information-processing systems, MNT sabotage agents would act on the real world. After damage, the world could not just be 'restarted from the original software' as a computer.

6.2.3 Criteria group III: protect humans, environment and society

Humans in peacetime could be endangered if military MNT developments became available to parts of civilian society. Nano-robots could be used by state agencies, enterprises or criminals to eavesdrop on people, influence them unknowingly, or for attacks on health and life. Unintended releases of self-replicating systems would usually be more dangerous in the military, since there damaging effects would have been built in. If molecular assembly became used widely, production of old or new weapons, including weapons of mass destruction, would only need access to the corresponding synthesis programs. Independent R&D and testing might be too complex for an individual, but a program made ready in the military and divulged to the outside could be set to action immediately. (Super-)human artificial-intelligence methods developed by the military to influence the thinking of an opponent's population might also be used to manipulate one's own people. Given all these dangers, MNT-based options for better protection against them do not count for much.

Dangers to the environment and sustainable development could ensue from releases of damaging agents developed in the military. Another problem may be overuse of resources consumed in a fast build-up of armaments. Grey goo would be the ultimate danger to the environment. Whereas in the civilian sector reluctance is probable, military R&D might nevertheless produce such aggressive nanosystems that could then be released intentionally or unintentionally. Also here, the dangers by far outweigh the potential for protection.

There are many different *dangers to the development of societal and political systems* that MNT could bring. Superhuman intelligences could take over power. The image of a human could be lost with cyborgs or subjects that change from normal life to an existence in software. MNT would provide many new possibilities for crimes that would increase societal instability. All of these should be discussed by society before they become feasible; finding appropriate regulation and arriving at agreement about it will certainly be difficult. Military R&D is obliged to find out all the negative potential uses of a new technology and make them practical, if possible; by doing so it could undermine a broad societal debate and create facts that might otherwise have been avoided.

Criminal misuse of MNT could bring many *dangers to the societal infrastructure*. Terrorist attacks could be directed against installations supplying goods or information services, and even a single nanosystem could, via self-replication, lead to a very large breakdown. Military R&D would create means for such attacks; diffusion to criminals could occur in the form of finished (maybe small, maybe replicating) systems or, with widespread molecular assemblers, even only the assembly programs.

6.3 Summary evaluation

Table 6.1 summarizes the evaluation under the criteria of preventive arms control of the various military NT applications presented in Section 4.1 and discussed in Section 6.1.[11] Whereas arms race and proliferation are tagged with a minus sign for nearly all table entries, dangers in the other criteria categories follow from more specific military applications. The following applications raise strong concerns:

- distributed sensors,
- new conventional weapons,
- implanted systems/body manipulation,
- armed autonomous systems,
- mini-/micro-robots with, but also without, weapon function,
- small satellites and launchers,
- new chemical and biological weapons.

It is in these areas where preventive limitations seem most urgent; they are discussed in Section 6.4.

In the area of nuclear weapons, computer modelling and, more so, very small nuclear weapons are linked with dangers. Limits are, however, not proposed in these two areas:

- Computer modelling would be practically impossible to limit and to verify, since ever smaller computers would be applied in ever higher numbers nearly everywhere. One can hope that, because the military would still insist that new warhead designs gained from modelling would be tested in reality before deployment, the existing ban on nuclear tests will suffice to prevent such innovations.[12]
- Very small nuclear weapons are still hypothetical. Thus, meaningful preventive limits would be difficult to define, and certainly difficult to agree about. In the related areas (antimatter storage, micro-ignition of nuclear fusion etc.) scientific-technical advances should be monitored closely. Design of limits should follow as soon as feasibility would come into view. A detailed study should be carried out soon.

There are two applications that show positive evaluations under several criteria: distributed sensors for treaty verification and chemical/biological

Table 6.1 Evaluation of potential military NT applications under the criteria of preventive arms control. Evaluation levels: – danger; 0 neutral; + improvement; u unclear; these refer only to the direct consequences. For an explanation of the criteria, see Section 5.2

	Effective arms control, disarmament and international law			*Maintain and improve stability*			*Protect humans, environment and society*			
	Arms control/ disarmament	*Humanitarian law*	*Weapons of mass destruction*	*Destabilization*	*Arms race*	*Proliferation*	*Humans*	*Environment/ sustainable development*	*Societal/ political systems*	*Infrastructure*
Electronics, photonics, magnetics	0	0	0	0	–	–	0	0	0	0
Computers, communication	0	0	0	–0	–	–	0	0	–0	0
Software/artificial intelligence	0	0	0	–0	–	–	0	0	–0	0
Materials	0	0	0	0	–	–	u	u	0	0
Energy sources, energy storage	0	0	0	0	–	–	0	0+	0	0
Propulsion	0	0	0	–0	–	–	0	0+	0	0
Vehicles	0	0	0	–0	–	–	0	0+	0	0
Propellants and explosives	0	0	0	0	–	–	u	u	0	0
Camouflage	0	0	0	0	–	–	0	0	0	0
Distributed sensors										
Generic	0	0	0	0	–0	–0	0	0	–	0+
Battlefield	0	0	0	–	–	–	0	0	–	0
Treaty verification	+	0	0+	+	+	+	0+	0+	0	0+
Armour, protection	0	0	0	0	–	–	0	0	0	0+

New conventional weapons										
Metal-less arms	−0	0	0	0	–	–	–	0	–	–
Small guidance	0	0+	0	–	–	–	0	0	0	–
Armour piercing	0	0	0	0	–	–	0	0	0	0
Small missiles	−0	−0+	0	–	–	–	–	0	0	–
Soldier systems	0	0	0	0	−0	−0	0	0	0	0
Implanted systems, body manipulation	−0	−0	0	−0	–	–	−0	0	–	0
Autonomous systems										
Unarmed	0	0	0	0	–	–	−0	0	0	0
Armed	−0	–	0	–	–	–	−0	0	0	−0
Mini-/micro-robots incl. bio-technical hybrids										
No weapon function	−0	0	0	–	–	–	0	0	–	–
Target beacon/armed	−0	–	0	–	–	–	−0	0	–	–
Small satellites/space launchers	−0	0	0	–	–	–	0	0	0	–
Nuclear weapons										
Auxiliary systems	0	0	−0+	0	−0	−0	0	0	0	0
Computer modelling	–	0	–	0	–	–	0	0	0	0
Very small weapons	–	−0	–	–	–	–	0	–	−0	−0
New chemical weapons	–	–	–	–	–	–	−0	−0	−0	–
New biological weapons	–	–	–	–	–	–	−0	−0	−0	–
Chemical/biological protection/ neutralization	0	0	+	−0+	−0	−0	+	0+	+	+

protection and neutralization. Here preventive limits would rather be counterproductive, however, this holds only if ambiguities are avoided:

- Sensors for verification should be designed in such a way that their use in battle would be difficult. This could be done, for example, by not making them mobile and requiring cable connections for power and communication. Also, the potential for their misuse for surveillance and tracking in civilian society should be minimized, e.g. by making them specific to treaty-relevant substances or events. For both purposes, the sensors should be kept to macroscopic size and not be made smaller than a few cm.
- Chemical/biological protection and neutralization, if coupled with an offensive programme, could even create destabilization and an arms race. Thus, offensive R&D should not be done. However, to a small extent, R&D of protection against potential new agents requires R&D of those very agents. The corresponding ambiguity should be minimized by guidelines, international transparency and a layered peer-review process for research that is relevant to the CWC and BTWC (as recommended by Steinbruner and Harris 2003 and Nixdorff 2003).

The cursory and preliminary evaluation of MNT under criteria of preventive arms control of Section 6.2 is summarized in Table 6.2. Here, the potential military applications have been grouped in five general categories. Autonomous, exponentially growing military production and MNT weapons would create nightmares in most of the criteria. Considerations on preventive limitations are given in Section 6.4.8.

6.4 Options for preventive limits on military NT

This section discusses how the most problematic applications identified in Section 6.3 could be limited preventively. Arguments for limits are balanced with potential positive uses, and aspects of effective but simple verification are considered.

6.4.1 Distributed sensors

Among distributed sensors, the strongest dangers for military stability as well as for privacy derive from those developed for use on the battlefield. As the size decreases, the risks would grow: small and very small systems would be difficult to see by eye or find by detector and could be used in high numbers. In addition, there is a certain ambiguity with respect to mini-/micro-robots for which a general ban is recommended in Section 6.4.5. On the one hand, there may be a motive to provide small sensors with at least rudimentary forms of mobility which could then be

Table 6.2 Summary evaluation of general categories of potential military applications of molecular NT under the criteria of preventive arms control. Evaluation levels as in Table 6.1

	Effective arms control, disarmament and international law			*Maintain and improve stability*			*Protect humans, environment and society*			
	Arms control/ disarmament	*Humanitarian law*	*Weapons of mass destruction*	*Destabilization*	*Arms race*	*Proliferation*	*Humans*	*Environment/ sustainable development*	*Societal/ political systems*	*Infrastructure*
Autonomous production using self-replication	–	0	–	0	–	–	–0	–	0	0
Sensing/communicating nano-robots	0+	0	0+	–	–	–	–0	0	–	–
Weapons	–	–0	–	–	–	–	–	–	0	–
Super-artificial intelligence	0	–0	0	–	–	–	–	0	–	0
Cyborgs etc.	–0	0	0	0	–	–	u	0	–	0

augmented over time. This would blur the distinction from small robots and undermine the prohibition of the latter. On the other hand, introduction of very small military systems would greatly aggravate the problems of verification.

Distributed sensors for general purposes (equipment monitoring, logistics, fixed-site security surveillance etc.) would be much less capable of doing general surveillance in a more or less autonomous mode. Often, they would be built into a larger piece of equipment, power for information transfer could be provided by wire or from an external, local radio-frequency field. Their potential misuse in societies would in many cases require significant new development efforts and could probably be contained by normal legislation and criminal prosecution.

As a consequence, limitation efforts should focus on distributed battlefield sensors. A total ban seems unrealistic, however. Several types of battlefield sensors are already deployed with armed forces – all of them of macroscopic size, sensor and relay units measure many centimetres (Blumrich 1998). In principle, many different characteristics could be used to define qualitative limits, such as the variable that is sensed, the mode of information transfer (e.g. only by cable, for radio transmission a maximum data rate) or a minimum allowed size. The former limits would be difficult to agree on and to verify. Because the dangers in most part derive from the smallness of NT-enabled new sensors, a simple size criterion seems most practical, combined with a capability of autonomous functioning and information transfer over some distance. As a consequence, one arrives at *a complete ban on self-contained sensor systems which are smaller than a certain size limit (3 to 5 cm)*. This ban should apply in the military as well as in the civilian sector and start at the development stage. It would not affect smaller sensors that are built into larger systems. One could also restrict the ban to include only sensors for outside use, but given the present emphasis on urban warfare, this would exclude an important part of military action and would introduce a problematic ambiguity.

At present, there are not many civilian needs for smaller sensors. Should an urgent need be felt, then exceptions need to be defined by clear qualitative and quantitative limits, with technical precautions against undermining of the main goal. An example is provided by the already existing medical intestine camera that is swallowed and transmits its video images only over about 1 m distance.

Some existing small monitoring and eavesdropping devices ('bugs', for insertion into telephone sets, sticking to the underside of tables etc.) would fall under the ban. In Germany, for example, their use is generally illegal; following the present proposal, production, import and possession should be forbidden, too. Small sensors used legally by law-enforcement agencies could be exempted. Use for espionage is of course illegal everywhere; the proposed ban will not be able to stop such covert use by intelligence services, but widespread application by armed forces can probably be prevented.

Because the lower size limit of centimetres is clearly macroscopic, verification could rely on on-site inspections and close visual observation of equipment to a large extent. Inspections would be directed to military installations, training and testing grounds. In order to be able to detect and investigate potential illegal sensor systems below 1 mm, portable microscopes should be allowed during inspections. Institutes and firms where development could take place need to be inspected, too, including research laboratories to check that they are not engaging in covert development efforts. In the civilian sector, inspection rights will be needed in industry and state agencies; depending on the exceptions, clinics and other institutions will have to be open to inspections, too.

6.4.2 New conventional weapons

Among potential new conventional weapons, the picture is varied. For armour piercing, no marked change is expected and thus limits are not recommended. For weapons and carriers that are similar to traditional armoured vehicles, artillery, combat aircraft and helicopters, counting them under the *CFE Treaty* if they fall under the respective definitions is obvious. If their parameters deviate, the definitions should be adapted and new ones introduced as required. Regulation similar to the CFE Treaty should be introduced in the other regions of the world. (For autonomous weapons/carriers, see Sections 6.4.4 and 6.4.5.)

Metal-less small arms and light weapons would pose strong dangers if they diffused to civilian society. Clear military advantages are not obvious, and arguments in favour of civilian uses do not come to mind. This suggests *a complete ban on small arms, light weapons and munitions that contain no metal.* The ban should hold for the military and civilian sector and start at the development stage. Should urgent reasons arise to develop such weapons and munitions, then producers should be obliged to include metallic patterns that immediately show up on x-ray and induction detectors. As long as small arms do not get smaller than about 10 cm and munitions about 1 cm, verification can rely on on-site inspections in military and security-force installations, testing and training grounds and production facilities. Visual observation should suffice, optimally supported by notification of types of arms and munitions and by individual markings on the arms.

Limits on small guidance and navigation systems would be extremely difficult to verify, because they concern mechanisms internal to the weapon/munition. To the extent that guidance/navigation systems contribute to a decrease in stability, it is probably more sensible to focus limits on the weapons or munitions containing them. In particular the problem of conventional attack against nuclear weapons and strategic command-and-control systems should be addressed.

Small missiles could destabilize the military situation if they are capable of precisely hitting sensitive points of strategic importance. More dangerous could be their use against persons by terrorists in civilian society. Compared to existing portable air-defence missiles and artillery rockets of 1.5 m length, small missiles below 0.5 m size would form a new class of its own, all the more so anti-personnel missiles below 5 cm size. Civilian uses are limited to already existing fireworks and life-saving apparatus for shooting a life-line to a stranded ship. As a consequence, *a complete ban on missiles below a certain size limit (0.2–0.5 m)* is recommended, with the exceptions mentioned. The ban should apply in the military and civilian sectors and start at the development stage. Verification is possible by visual observation during on-site inspections in installations of the military and security forces, including testing and training grounds, and in production facilities.

6.4.3 Implanted systems/body manipulation

Body implants are already now playing an important role in medical diagnosis and therapy. Artificial hip joints are purely mechanical; heart pacemakers have sensors, processing electronics and stimulating electrodes. The role of sophisticated implants will increase with advances in MST and in particular NT. Other types of body manipulation can, for example, use drugs that target specific organs or modify metabolism.

One class of future implants would monitor the biochemistry and release therapeutic drugs as required, e.g. sense the blood-sugar level in diabetes patients with immediate release of the correct amounts of insulin. Similar principles will likely be applicable to many diseases, e.g. chronic pain, schizophrenia and dementia. Medically motivated R&D for such purposes, and the eventual application, should of course not at all be hampered by efforts to prevent certain military uses of implants.

However, even in the civilian sector problems would arise when drug-releasing implants were used for preventive application without illness, or for improving the human, e.g. for better memory or learning, stronger muscles, longer periods of wakefulness. More serious problems would ensue from implants that release mood-controlling drugs, such as creating feelings of happiness. Here a danger of addiction has to be feared, even if damaging secondary effects could be avoided by better understanding of the biochemical interactions.

Another class of medical implants makes contact to nerves or the brain, picking up signals electrically or electrochemically and/or stimulating muscles, nerves or the brain. Already at present auditory prostheses are implanted routinely in the cochlea of the inner ear, overcoming some types of deafness by electrically stimulating the hair cells. Research is underway for implants into the eye that, by electrical stimulation of the retina, could restore some vision in certain kinds of blindness. In epilepsy research, multi-electrode arrays on or in the brain cortex are used to find foci of seizures; chronic implantation could serve to detect and suppress an impending seizure. Similar brain electrodes are used to suppress tremor in Parkinson's disease. To restore motion in patients paralysed by injury of the spinal cord, signals produced by the intent to move a limb could be sensed in the motor cortex, processed electronically and fed to the efferent nerve below the injury, or the muscles could be stimulated directly. In case of a lost limb, these signals could control actuators in a non-rigid prosthesis.

Such medical use to partly restore lost body functions is ethically not controversial. For the time being, sensory or motion prostheses will be inferior to the natural organs. The problems will come when artificial organs can provide better capabilities than the natural ones, or when brain-machine interfaces can transmit complex sensory impressions or general thoughts. Such implants would clearly represent a qualitative step with an opportunity for preventive limitation.

Step by step, medical progress will likely approach such a stage, probably with accelerating speed by advances in many fields. Medical ethics will be refined along the way. However, the issue of body modification for other purposes than restoring health will pose new, fundamental questions. One can easily point out potential benefits. Artificial bones could increase strength and avoid osteosclerosis; an implanted computer with a brain interface capable of transferring complex information would allow direct connection to the world-wide communication network and information resources. It is equally easy to conceive of misuse, e.g. by the chip producers or providers of downloadable software. Similarly to the handling of the issues of genetic manipulation and improvement of humans, society should carry out a thorough debate about limits to body manipulation and enhancement by implants that release drugs or provide brain-machine contact, by artificial organs or modified biochemistry.

This debate will require considerable time and maybe experiences with the first tests of non-medical manipulation. It should not be undermined by developments in the armed forces. Because, in civilian medicine, ethical considerations are routine when new technologies are being tested and introduced, one could conceive of a prohibition of non-medical body manipulation in the armed forces. However, this is problematic for two reasons. First, there may be other areas where motives for such implants could arise, maybe fire fighters, police or emergency pilots. Second, civilian society might arrive at a decision that would allow certain implants under certain conditions, and then a military prohibition would no longer be tenable. Since the main goal is to provide time for society to arrive at a sound general decision, it seems best to introduce a temporary prohibition that covers all cases, civilian and military. Thus one arrives at the demand for a *moratorium on implants and other body manipulation that are not directly medically motivated.* The moratorium should *comprise the civilian and military sectors and last for 10 years*, with the possibility of prolongation.[13] It should be agreed upon by international understanding, involving bodies responsible for international co-ordination in the medical field such as the World Health Organization (WHO). Should verification of compliance be deemed necessary, the WHO might conceivably play a role. For challenge inspections access to medical facilities, surgery records etc. together with rights to interview medical personnel as well as people who might have been subject to implant surgery or other body manipulation would be needed.

Certainly here as elsewhere there will be a grey area and the need to agree on interpretations. To give a few examples: devices worn on the skin that sample body fluids and release drugs through micro-needles would not count as implants. Rules would be needed on how to deal with people who volunteer for getting certain implants. Chip implants with medical data to be read out for emergency use seem innocent at first, but could serve to open a Pandora's box of subsequent less innocent applications, not only in the military sector. Body manipulation achieved with drugs that are transported into the body by other means than implants represent a large grey area, starting maybe with drinking coffee or with alertness-increasing amphetamines. Agreed limitations can take into account the kind and strength of the effect, and the laws about addictive drugs and doping in sports can form a precedent.

It is to be hoped that despite different national cultures, agreement on fundamental approaches can be achieved in the appropriate international bodies.

6.4.4 Armed autonomous systems

Armed robotic systems that would move on the ground, on or under water, or in the air need yet to be studied systematically under the aspects of preventive arms control. The problems they pose are mostly independent of NT, but NT-based computing power will be needed to achieve the required autonomy. NT can also come in with many types of sensors and actuators. This section presents first considerations on limiting options.[14]

There are practically no civilian uses of armed robots. One conceivable exception might be robots substituting human guards in securing closed facilities against intruders. Here small arms or non-lethal weapons would suffice, but these could already raise objections under constitutional criteria. Another exception would be robots used by the police to fight against hostage-takers. If accepted at all, such robots should be equipped only with arms similar to those of the police, and the mobility and number should be limited. Thus, they would be impractical for military combat and would not create ambiguity. The already existing remotely controlled robots for manipulating and shooting at suspected bomb briefcases fulfil these conditions well.

The military have considerable interest in armed autonomous systems – the combat efficiency would be higher due to higher agility and smaller size; fewer soldiers would be injured and killed; operations might be cheaper. However, these views are secondary considering the generally negative evaluation under the preventive-arms-control criteria in Section 6.1. Because armed autonomous systems would present a clear qualitative step, a *general prohibition of re-usable armed, mobile systems without crew* is possible and probably represents the best solution. Most armed forces have not yet introduced such systems; the USA would have to remove the option of mounting a Hellfire missile on its Predator drones.

The wording 'without crew' instead of 'autonomous' has been chosen to include remotely controlled systems. Even though these are not autonomous in the proper sense of the word, they could be deployed in high numbers and later relatively simply and covertly changed to autonomous operation. Also, verification of autonomy versus remote control would be very difficult from outside.

Cruise missiles are already introduced with many armed forces. In one sense, they are an armed system that autonomously approaches a target and attacks it. In another sense, one could argue that a cruise missile follows a pre-planned course to a target and explodes there (one-time use), similarly to a ballistic missile or an artillery grenade. While there are reasons to argue for a ban on (certain types of) cruise missiles,[15] they are different enough from the systems discussed here that one should not burden the prohibition of new systems with the demand to withdraw and dismantle others that are already widely introduced. The proposed term 're-usable' also excludes target-seeking sub-munitions for which similar arguments hold.

With a ban on armed systems, robots without weapons must not create opportunities for circumvention. For many autonomous systems of sizes from 1 m (ground, air) to 100 m and more (sea) one could already discern weapons by external observation from some distance – a ground vehicle carrying a gun and a turret versus a lorry, an aircraft provided with missile-mounting pods or bomb hatches versus a photo-reconnaissance plane, a surface ship having guns or missile launchers versus a cargo boat. New unarmed systems such as a ground-mobile robot for demining or rescue of injured soldiers should be designed clearly without weapons function. However, such systems will often be modular, so that a manipulator, for example, could later be exchanged by a firearm. Here numerical limits and rules of warfare – as with medical vehicles under Additional Protocol I to the Geneva Conventions of 1949 (ICRC 1977) – could help.

Extending the ban to include all autonomous military systems would remove the ambiguities about unarmed systems. However, the first robot lawnmowers and vacuum cleaners are already on the market, and the use of service robots of different kinds in civilian life is expected to increase rapidly over the coming years (on questions of civilian liability etc. see Christaller *et al.* 2001: Ch. 5). Under such circumstances it would be unrealistic to exclude military use of similar systems, for example, for logistics and transport. Robot control of weapons, on the other hand, crosses an obvious threshold, as was also recognized by a military author:

> Beyond technological obstacles, the potential for effective battlefield robots raises a whole series of strategic, operational, and ethical issues, particularly when or if robots change from being lifters to killers. The idea of a killing system without direct human control is frightening. Because of this, developing the 'rules of engagement' for robotic warfare is likely to be extaordinarily contentious. How much autonomy should robots have to engage targets? . . . Should the United States attempt to control the proliferation of military robotic technology? Is that even feasible since most of the evolution of robotic technology, like information technology in general, will take place in the private sector? Should a fully roboticized force be the ultimate objective?
>
> (Metz 2000; see also Section 1.5.7)

For answering the question of limits on unarmed robotic systems, one can look at the Treaty on Conventional Forces in Europe (CFE Treaty) of 1990, amended and updated 1992 and 1996 (CFE 1990; the 1992 CFE-1A agreement introduced numerical limits on personnel). Unfortunately, similar agreements are not yet even in discussion, either for other regions of the world or for naval forces. In order to achieve 'a secure and stable balance' and to eliminate 'the capability for launching surprise attack and for initiating large-scale offensive action in Europe', the Treaty limits main weapons of the armies in five categories – battle tanks, armoured combat vehicles, artillery, combat aircraft and attack helicopters – and regulates equipment in six categories – primary trainer aircraft, unarmed trainer aircraft, unarmed transport helicopters, armoured vehicle launched bridges, armoured personnel carrier look-alikes and armoured infantry fighting vehicle look-alikes. Transport lorries or transport aircraft are not limited.

Pursuing and extending this logic, robotic systems clearly without combat function, such as robotic lorries or pilot-less transport aircraft, might be left unregulated. However, in order to reduce the potential for arming them, numerical limits seem advisable also here. States could introduce different categories of such systems – transport, search and rescue, demining etc. – and agree on respective maximum holdings. Crewless military land, sea and air vehicles would raise problems for traffic safety already in peacetime, particular dangers can be foreseen within cities and in congested air space (e.g. all over West and Central Europe). As a consequence, a *ban on such systems in public areas and air space* seems advisable, limiting them to military premises (barracks, testing or training grounds). If civilian society had accepted civilian autonomous vehicles/robots and had introduced appropriate rules for their movement and behaviour, then similar regulation might allow some military systems as well.

It may turn out that a comprehensive prohibition of armed autonomous systems will not be politically feasible. In such a case, the least one should demand is preserving the CFE Treaty. Its definitions of the armaments do not mention crews, e.g.:

> (C) The term 'battle tank' means a self-propelled armoured fighting vehicle, capable of heavy firepower, primarily of a high muzzle velocity direct fire main gun necessary to engage armoured and other targets, with high cross-country mobility, with a high level of self-protection, and which is not designed and equipped primarily to transport combat troops. Such armoured vehicles serve as the principal weapon system of ground-force tank and other armoured formations. Battle tanks are tracked armoured fighting vehicles which weigh at least 16.5 metric tonnes unladen weight and which are armed with a 360-degree traverse gun of at least 75 millimeters calibre. In addition, any wheeled armoured fighting vehicles entering into service which meet all the other criteria stated above shall also be deemed battle tanks.
>
> . . .
>
> (F) The term 'artillery' means large calibre systems capable of engaging ground targets by delivering primarily indirect fire. Such artillery systems provide the essential indirect fire support to combined arms formations.
>
> Large calibre artillery systems are guns, howitzers, artillery pieces combining the characteristics of guns and howitzers, mortars and multiple launch rocket systems with a calibre of 100 millimeters and above. In addition, any future large calibre direct fire system which has a secondary effective indirect fire capability shall be counted against the artillery ceilings.
>
> . . .
>
> (K) The term 'combat aircraft' means a fixed-wing or variable-geometry wing aircraft armed and equipped to engage targets by employing guided missiles, unguided rockets, bombs, guns, cannons, or other weapons of destruction, as well as any model or

version of such an aircraft which performs other military functions such as reconnaissance or electronic warfare.

The term 'combat aircraft' does not include primary trainer aircraft.

(CFE 1990: Art. II)

Robotic vehicles similar to the traditional ones would immediately fall under these definitions, and would thus count against the Treaty limits.[16] For land vehicles of lower weight or calibre, the CFE categories of heavy armament combat vehicles, armoured infantry fighting vehicles and armoured personnel carriers could apply.[17] Potential new types of vehicles/robots might fall outside of treaty definitions, such as a battle tank below 16.5 metric tons equipped with an electromagnetic gun of less than 75 mm calibre, or person-size walking combat robots. To close such loopholes, additional categories with appropriate limits will be needed.

For continuing effectiveness, agreements similar to the European Treaty would be needed on a global scale.

If autonomous combat vehicles cannot be prevented in general, crewless carriers could also be used for nuclear weapons (bombers, maybe also submarines, land-mobile intercontinental ballistic missiles or artillery). The least one should demand in such a case is *that no qualitatively new types of nuclear-weapon carriers* be introduced, e.g. a walking robot carrying a bomb. Reasons of safety and security will anyway argue for continuing direct human control of nuclear weapons.

Assuming that autonomous combat systems could not be prevented, the question arises whether autonomy should include the decision to direct and use a weapon against a specific target. Because autonomy in general brings destabilization and because at least the first generations could not guarantee compliance with the international humanitarian law, *the weapon targeting and release decision should remain in human hands*, that is, it should only be done using remote control. It has to be admitted that such a rule would be difficult to be upheld in an environment full of crewless systems mutually observing and threatening each other. This is another argument for strict numerical limits, including small systems (see Section 6.4.5).

Neither the proposed strong prohibition nor the mentioned weaker solution would restrict development and use of general-purpose robots in the civilian sector, so that here no balancing of benefits and costs is needed. The only exception would be armed systems for law-enforcement purposes – in order to prevent military build-up in the guise of police systems, some restraint will be needed. This is no serious problem, however, since civilian society will have its own motives against too strong mechanization of law enforcement.

Verification of the proposed ban on crewless armed, mobile systems could be based on external observation during on-site inspections and manoeuvre observations. Due to the size of these systems, weapons, weapons pods etc. could easily be discerned from the outside at short distance (a few metres). For the same reason, national technical means of verification could play a significant role for those who own them. As with the CFE Treaty (and others since 1987), regular data exchanges on holdings and armaments properties would form the precondition that random and challenge inspections need be sent only to a small sample of the installations in question. To include development and testing, data exchanges and inspection rights should extend to development projects and testing areas.

Verification of the ban on autonomous weapon release and the obligation to use remote control by a human operator is practically not feasible, because the software governing release could be easily modified (verification of embedded software would be prohibitively intrusive anyway). This is no strong obstacle, however, because most rules of the international law of warfare cannot be verified in advance. They rather form norms for behaviour that can be evaluated after the action.

6.4.5 Mini-/micro-robots

While some considerations of the preceding section apply also to small robots, the latter pose additional problems: they could be used in very large numbers and they could be easily hidden or would even be invisible. Because small robots (including bio-technical hybrids) with and without weapons function would bring strong dangers under military as well as civilian viewpoints, they should be prevented as completely as possible, taking into account indispensable positive uses. Small robots would represent qualitatively new systems, thus preventive limitation is possible.

For a clear definition and in the interest of verification, neither details of the inner working (such as NT-based control computers, sensors or actuators) nor the degree of autonomy nor the presence of a weapon mechanism should be considered. Simple characteristics visible from the outside are mobility and size. Some mobility mechanisms (legs, wheels, tracks) would be immediately obvious. Others could be hidden inside the body, but could be observed in operation, e.g. a single rod that is only explosively protruded from the 'belly' when the system starts a hop. Very small systems would need examination under some type of microscope. Crewless weapons and systems that are introduced already with the armed forces are above 1 m size: on land, there are at most prototypes of robots; in water, torpedoes and a few deep-diving robots; in the air, cruise missiles and uninhabited aerial vehicles/drones. To avoid demanding removal of existing armaments, a lower size limit below 1 m is advisable, somewhere between 0.5 and 0.2 m. This would be far enough above microscopic size that observation by eye would normally suffice during inspections; a covert technological jump from 0.5 m to less than a millimetre is improbable for fully artificial systems. Nevertheless, as soon as microscopic mobile systems become possible in the civilian sector, inspectors would need the right to use microscopes. Because small insects could be implanted much sooner, this right is required from the outset.

The military interests in small robots for combat support and actual combat might turn out less than expected due to their limits in speed, range and payload. In any case, the dangers far outweigh the military advantages. Surveillance from the air, for example, can more effectively be done using larger craft, such as the existing uninhabited aerial vehicles. Mine detection and removal as well as rescue of injured soldiers could also be carried out better with systems above 0.5 or even 1 m size: the payload and range would be higher.

Mini-/micro-robots could be used beneficially in civilian society (see also Altmann 2001: Section 7.2 and refs). For several potential uses on/in land, water or air, however, systems above 0.5 or 1 m size would be suited better; examples are:

- remote characterization of disaster areas,
- monitoring of pollutants,
- replacing weather balloons,
- tracking populations of endangered species.

For other applications, fixed sensors could do the job, e.g.

- monitoring of traffic,
- finding activity connected with illegal border crossing.

In law enforcement, applications for

- covert search for contraband,
- covert monitoring of hostage situations,
- monitoring of criminal suspects

may seem justified, but raise significant problems with privacy and human rights, in particular if the mini-/micro-robots are smaller than 1 mm and are produced in high numbers. Here, prevention of misuse will demand either a total renunciation or very strict qualitative and quantitative limits.

Among the remaining conceivable civilian applications there are only very few where mini-/micro-robots below 0.2 or 0.5 m size could not be replaced by larger ones (the optimum size range is given in parentheses):

- exploration of the moon, planets and asteroids (10–50 cm),
- exploration of shattered buildings after an explosion or earthquake (5–20 cm),
- investigation of the interior of a dangerous building (fire, toxic substance etc.) (5–20 cm),
- inspection of narrow pipes (0.5–10 cm),
- surgical robots for movement and operations in the body (0.1–1 mm),
- toys (5–50 cm),
- amateur objects (<1 mm–50 cm).

For exploration of celestial bodies, only very few copies would be needed, and there would be a motive to keep them lightweight and thus fragile; it would not be difficult to formulate an exception for single robots. Swarms of centimetre size flying or crawling robots for moon and planet investigation would present a harder problem; cheap production of hundreds or thousands could lead to diffusion to uses on Earth. For investigations in shattered or dangerous buildings, systems below about 5 cm would be less than optimal because they could be blown away by flames or other draughts. By exception, limited numbers of 5–20 cm size robots could be allowed for fire fighters and other rescue personnel. For pipe inspection, neither general mobility nor autonomy is needed; misuse of robots made for this purpose could be prevented by restricting their capabilities. Similar considerations hold for surgical robots in the body; here, military medicine would be affected similarly and could be included. Fully capable small robots used as toys for children would bring a great potential for accidents and abuse. Thus, toy robots should be clearly limited – in mobility, sensing/actuating capability or autonomy. To hamper certain covert uses, they should not be produced below about 20 cm size. Existing robotic 'pets' and dolls are in this category; the toy industry will probably not massively oppose such limitations that society would introduce to protect higher values. More difficult could be amateurs, hobbyists and student teams who would like to remain up-to-date with their model aircraft, small robots etc. Robotic soccer games use 'players' of size 0.1 to 1.2 m, there are competitions in aerial robotics and micro-air vehicles.[18] Already now certain limitations are in effect – by law or as a matter of course for responsible behaviour – e.g. concerning not using model aircraft in cities or near airports. As soon as more capable small land/water/air robots become possible, society will have strong motives to regulate them. To avoid ambiguities, no exception for amateurs should be made – they should not be allowed to use robots capable of moving outside of their homes or laboratories that are smaller than 0.2 or 0.5 m size. Certainly the majority of amateurs would accept such a rule and comply with it. Those very few who might not comply would need to be prosecuted and fined, as usual with other laws.

Except for medical surgery as a temporary action, there seems to be no important civilian application of robots below 1 mm size for the next two decades. Micro- and nano-robots to stay in the body would pose different problems; molecule-size systems used as military weapons are banned by the Chemical Weapons Convention (see Section 6.4.7). Civilian application would need extremely strict rules to prevent abuse. Because this is still hypothetical, no specific limits are considered here.

As a consequence, one arrives at a *general prohibition of small mobile (partly) artificial systems*[19] *below a certain size limit (0.2 to 0.5 m) in all media, in the military and the civilian sector*, independent of the degree of autonomy and the biological-technical mix.[20] The prohibition should include development and testing. By default, it includes systems down to arbitrarily small sizes. If this rule is accepted, use of mini-/micro-robots as surrogates for anti-personnel mines is excluded simultaneously.[21]

Exceptions should be strictly defined and narrowly limited; in the civilian sector, they could concern exploration of celestial bodies, exploration of shattered or dangerous buildings, inspection of pipes and surgical operations. No exception should be made for toys and amateur robots. In the military sector, the only exception recommended is for surgical operations.

To make sure that the exceptions are not used to circumvent the purpose of the ban – namely, securing military stability and preventing civilian misuse – special technical limitations as well as licensing procedures should be used. Such measures include:

- reduced autonomy – e.g. by requiring remote control most of the time, with only an emergency capability to function during periods when communication is lost,
- reduced range – in many cases, more than 100 m from the operator is not required,
- limit on speed,
- lower limit on size,
- external power supply with only little reserve capacity on board (e.g. provided by cable, induction loop or microwave guided in pipe),
- purpose of the robots, organization that would use them,
- maximum number of allowed robots.

These measures should be combined as appropriate for the respective purpose.

Verification of compliance with the general prohibition and the exceptions can to a large extent rely on on-site inspections in military installations and training grounds. Here, the task is detecting possible illegal mobile systems below the size limit. Visual close inspection of equipment and observation of its use will largely suffice as long as robot systems below about 1 mm are not practical. To provide a chance for their detection already from the beginning, portable microscopes (at least optical, maybe also of the scanning-probe type) should be allowed during inspections. Molecule-specific sensors, including for detection of genetic modification, would probably rather count as inspection equipment under the Chemical or Biological Weapons Convention. To find evidence of illegal development and testing, respective institutions need to be inspected. Even though no limit on research is proposed, research laboratories should be included since development and testing could be carried out here, too.

Because nearly all exceptions are foreseen in the civilian sector, inspection rights are required here as well, mainly directed to industry, but also to law-enforcement agencies. To make sure that illegal manipulations of insects and other small animals can be detected, there needs to be a right to inspections of biological laboratories. Observation of public life and evaluation of public media will strengthen the verification effectiveness.

6.4.6 Small satellites and launchers

Small satellites would bring dangers due to their possible use as space weapons.[22] Banning satellites smaller than a certain size would exclude that option, but would prevent positive uses of small satellites, and would not restrict larger space weapons. Because there are no space weapons yet,[23] the qualitative step applies still to the general category, not only the small sort. The best solution for avoiding destabilizing developments in space would be a *comprehensive ban on space weapons of all kinds, including development and testing.*[24] This would allow all kinds of civilian small satellites, and military ones for non-weapons purposes. The risk of their misuse for anti-satellite attacks could be minimized by strict numerical limits, together with rules of the road that forbid close approach to other than one's own satellites.

Verification could to a very large part rely on space observation from Earth by radar and optical cameras, optimally augmented by pre-launch inspection after notification; space-to-space observation would add some capability.

An outright ban on small satellites would provide better security, but would forego the positive uses. If the latter are deemed important enough, the space-faring nations should agree – optimally in the UN framework – on detailed rules for the numbers, orbits and launches of small satellites, together with an international process of notification, licensing and pre-launch inspection. Otherwise, suspicion and instability would rise.[25]

Small space launchers should be included in the above regulation; they should only be allowed for legitimate small satellites. In order to minimize ambiguities with other small missiles (see also Section 6.4.2), challenge inspections of missile test ranges and military and civilian space-launch facilities should be agreed upon.

6.4.7 New chemical and biological weapons

New chemical or biological weapons enabled by NT would clearly violate the Chemical Weapons Convention (CWC) or the Biological and Toxin Weapons Convention (BTWC) (CWC 1993; BTWC 1972). In so far as there are practically no such weapons known at present, new types would represent a qualitative step. The recommended option thus is to *uphold both Conventions and reinforce their universal scope.*

For the CWC, advances in molecular-biological understanding of damage mechanisms and in synthesis of potential new agents need to be followed up to expand, as required and foreseen in the CWC, the Schedules 1–3 in the Annex on Chemicals. The functional definitions in Art. II are applicable and need not be changed:

> 1. 'Chemical Weapons' means the following, together or separately:
> (a) Toxic chemicals and their precursors, except where intended for purposes not prohibited under this Convention, as long as the types and quantities are consistent with such purposes;
> (b) Munitions and devices, specifically designed to cause death or other harm through the toxic properties of those toxic chemicals specified in subparagraph (a), which would be released as a result of the employment of such munitions and devices;
> (c) Any equipment specifically designed for use directly in connection with the employment of munitions and devices specified in subparagraph (b).
>
> 2. 'Toxic Chemical' means:
> Any chemical which through its chemical action on life processes can cause death, temporary incapacitation or permanent harm to humans or animals. This includes all such chemicals, regardless of their origin or of their method of production, and regardless of whether they are produced in facilities, in munitions or elsewhere. (For the purpose of implementing this Convention, toxic chemicals which have been identified for the application of verification measures are listed in Schedules contained in the Annex on Chemicals.)
>
> 3. 'Precursor' means: Any chemical reactant which takes part at any stage in the production by whatever method of a toxic chemical. This includes any key component of a binary or multicomponent chemical system.

> (For the purpose of implementing this Convention, precursors which have been identified for the application of verification measures are listed in Schedules contained in the Annex on Chemicals.)
>
> (CWC 1993: Art. II)

Paragraph 1.(a) contains a general-purpose criterion. Note that toxicity in Paragraph 2. is related to humans and animals; chemicals against plants or equipment are not covered by the CWC. New NT-enabled agents are included if they are a 'chemical' and damage by 'chemical action'. Some doubts could arise with supramolecular systems such as nanomachines that would act mechanically, electrically or thermally on cell components. To avoid undermining the CWC, a clarifying interpretation would be needed, e.g. stating that all interactions of agents smaller than cells count as chemical. (Such an interpretation could be concluded at the next Review Conference in 2008.)

Also for new substances there is the unsolved question: what exactly is a riot control agent and what is law enforcement where the former can be used (whereas they are forbidden as a method of warfare)?

The verification means and methods that the CWC contains in its Annex on Implementation and Verification – declarations, inspections, equipment, seals, sensors, sample-handling, facility agreements etc. – seem to be sufficient for new agents.

The BTWC is more comprehensive in its scope, defined in Art. I:

> Each State Party to this Convention undertakes never in any circumstances to develop, produce, stockpile or otherwise acquire or retain:
> (1) Microbial or other biological agents, or toxins whatever their origin or method of production, of types and in quantities that have no justification for prophylactic, protective or other peaceful purposes;
> (2) Weapons, equipment or means of delivery designed to use such agents or toxins for hostile purposes or in armed conflict.
>
> (BTWC 1972: Art. I)

Also the BTWC uses the general-purpose criterion in paragraph (1). However, there are no restricting further definitions, thus this convention includes not only agents against humans and animals, but also against plants or equipment (and there is no exception for riot-control agents). New biological agents and toxins created with the help of NT are covered. Doubts may arise with respect to biological-technical hybrid organisms, 'artificial microbes' and nanomachines that are clearly non-biological. To prevent endangering the Convention, at least an additional clarification/interpretation would be needed; one solution could be to state that the Convention covers also hybrid and artificial systems that could enter an organism. (A clarifying interpretation could be agreed upon at the next Review Conference in 2006.) The more systematic solution, a change in wording, might be more difficult to achieve. Toxic substances that are not of biological origin or are not produced by biological systems would not count as toxins, but would fall under the CWC, i.e. be prohibited if directed against humans or animals.

The main weakness of the BTWC is the lack of verification; this problem has been recognized for a long time and becomes ever more urgent in the light of biotechnological advances. NT will accelerate such advances. The work on a compliance and verification protocol started more than a decade ago, but has been stopped due to resistance mainly from the present US government (Nixdorff *et al.* 2003: Chs 8–11 and refs). In order to strengthen the BTWC, a new process towards a compliance and verification protocol should be started and brought to a fast conclusion. The existing rolling text with its provisions for challenge inspections in pharmaceutical laboratories, managed access, sample-taking etc. (BWC AHG 2001) has been designed with present biotechnology in mind, but could probably apply to NT-based further advances.

As in other areas, fastest progress in biotechnology is expected in civilian R&D and industry. This opens many possibilities for misuse, including exploitation of differences in national legislation. To minimize the risks, states should conclude a *Biosecurity Convention* that should be co-ordinated closely with the BTWC and its future verification and compliance mechanism (e.g. Barletta *et al.* 2002; Sunshine 2003; Tucker 2003).

Beyond the existing overlap concerning toxins, NT will bring additional dimensions of blurring between chemical and biological interactions and, above that, also between biological and artificial systems. From a scientific viewpoint, new technologies that are not sufficiently covered by existing arms-limitation treaties demand new agreements. Not only the weapons can be new, but also the targets, e.g. robotic systems, bio-technical hybrids. For a systematic approach, one would want to complement the CWC and the BTWC by a third convention covering hostile interactions within the body or cells of living systems by artificial microscopic agents that may have entered by whatever pathway, through the skin, by inhalation, ingestion or injection, via nanometre-size pores in cell membranes. This is similar to the proposal of a treaty prohibiting the hostile manipulation of human physiology (Wheelis 2003). It would also be possible to include traditional chemical and biological/toxin agents into a new, comprehensive convention that would supersede the CWC and BTWC. (This could then close the gap concerning anti-plant (and maybe anti-matériel) agents in the CWC, and concerning verification in the BTWC.)

Looked at pragmatically, however, it would probably be more complicated politically to negotiate a new convention with all the ensuing discussions about scope and definitions; for the time being, adding interpretations to the existing CWC and BTWC is certainly easier. And opening up these Conventions for new discussion could also result in weakening them. Thus, keeping them intact and widening their scope to include NT-based new possibilities seems the best approach for the near and medium future.

6.4.8 Molecular NT

MNT is still hypothetical. It would represent a clear qualitative step so that preventive limitation would not need removal of systems that already exist. Should MNT become feasible, it would be of the highest importance to prevent all the potential dangers mentioned in Section 6.2, such as the breakdown of arms control and disarmament, all sorts of new weapons that could even 'grow' on site via self-replication, unprecedented destabilization, a potential for utmost destruction even in the hands of single individuals. Containing the risks would already be difficult within civilian societies.[26]

If the international system remains without an authority comparable to the one of the state within societies, so that threats from armed forces or state-tolerated international terrorism persist, preventive limitation would be much more difficult. This is mainly because the degree of intrusiveness needed to adequately verify compliance would have to be similar to that in civilian society, but this would intrude so deeply into military secrecy that it could endanger national-security interests – at least if worst-case assumptions are made. On the other hand, the alternative of unlimited qualitative and quantitative arms build-up would be so dangerous that overriding the military hesitations by high-level political decision seems at least in principle possible.

Verification of international obligations not to develop certain types of universal molecular assemblers, programs for them, human-like artificial entities etc. would certainly mean very stringent requirements, but should be possible if the same methods as within societies are applied. Whoever doubts this should have the same doubts on the efficacy of regulation within states.[27] Also inside states there may be competing firms, state agencies, criminal groups who could develop and use MNT for dangerous, illegal purposes.

It is obvious that effective regulation of MNT will need to cover both civilian and military aspects, on the national as well as international level. Because MNT and its associated concepts would bring fundamental changes, they pose fundamental questions:

- Is it possible to steer these technologies away from disastrous uses, or would humankind be served better by B. Joy's proposed relinquishment – not developing certain types of genetics, NT and robotics, limiting our pursuit of certain kinds of knowledge (see Section 1.5.5)?

One can have sympathy for such a restrictive approach, especially because there are already precedents concerning manipulation of human embryos etc. The counter-argument is that science and technology are driven by too powerful factors. Even if that were the case, however, one need at least not rush forward as fast as possible – R&D spending does not follow from the laws of nature, but is decided by political bodies according to political considerations. In particular technologically leading states can afford a slower pace; part of their responsibility is to act as role models and to apply the precautionary principle while it still can have an effect.

- Could international regulation and verification of military uses of MNT be agreed upon on a similar level as required within civilian society?

This seems difficult to conceive as long as national security is still perceived as deriving from the threat of armed forces.

- Turning this argument around: could the advent of powerful new technologies be instrumental in raising the global order to a new level, more similar to the organization within civilized democratic states (see also Section 7.4)?

Throughout history, technological waves have been linked with changes in societal organization and state behaviour. Already existing technologies, in particular the information technologies and the Internet, are acting in the direction of a global society with global governance, or could at least be used to further that purpose. Advances in NT, biotechnology, information technology – which will arrive before MNT could appear – could be used to get states and peoples accustomed to the idea of stronger international co-operation with legislation, criminal investigation and prosecution.

Because the feasibility of MNT is open, details of its arrival are not available and preventive limits cannot yet be designed. As soon as MNT turns out practicable, there would be very strong economic as well as military pressures to advance fast. In such a case, developments should not overtake limitation negotiations – there should already be available systematic and reliable considerations about preventive limits for MNT. An alleviating factor is that MNT would pose fundamental questions for civilian society so that strong motives for general technology assessment and preventive regulation would exist. Seen from today, it seems that limits on MNT would need to be enacted early, comprising practically all areas of application in the civilian and military sectors. Even small loopholes could undermine the limits.

To enable responsible decisions early, at present it seems most important to carry out studies of:

- MNT feasibility,
- the potential time frames of introduction of various applications,
- military applications and their dangers,
- military–civilian interactions,
- appropriate preventive civilian and military limits,
- methods and technical means of verification in both sectors.

Such work should begin soon; of course the first two topics are crucial for the latter ones.

6.5 Meta-aspects concerning preventive arms control

In writing this chapter, several considerations about preventive arms control on a meta-level came up that warrant future discussion. They are presented in the following.

1. The existing set of criteria as given in Section 5.2 is roughly appropriate, but some refinement may be useful:
   - The arms-race criterion (II.2) is fulfilled with nearly any military application. This could just be accepted as mirroring developments to be expected in practice, or the criterion could be sharpened to cover only exceptionally strong arms-race motives. Obviously all technologies that make the military more efficient in offence or defence will provide arms-race motives; this is less so for indirect, auxiliary functions such as better injury treatment. The criterion does not seem superfluous between opponents of roughly equal capability that want to avoid war, because simply accepting an arms race would lead to a similar situation, only on a qualitatively higher level of armament. Even if no destabilization ensued, at least money could be saved by mutually renouncing some new types of armaments. On the other hand, some technological advance cannot be avoided – the armed forces cannot be held artificially at a lower technological level than their societies.
   - Arguments along similar lines could be made concerning the criterion on proliferation (II.3).
   - Separating the criteria about humans and environment/sustainability has worked well.
2. The strongest dangers were found, not unexpectedly, with offensive weapons. Should the criteria in some way be sharpened to better reflect a bonus on defensive structures and equipment, or is this already built in sufficiently?

3 The evaluation using three levels (−, 0, +) is very coarse and cannot cover smaller differences. Could a more differentiated scheme be devised, or would it just pretend to be more exact because the prospective judgement is by necessity only qualitative?
4 The discussion on manipulation of soldiers' bodies brings up the question of human rights for soldiers. Should there be a right for them to decline inoculations, biochemical manipulation, implants etc.? Should such a right be agreed upon internationally?
5 Even though NT will make possible new systems and missions, consideration of comprehensiveness and verifiability led to limitation proposals that do not focus on NT as such, but rather on specific classes of systems and missions, independent of the technology used.
6 Several of the recommendations are not NT-specific, but relate to arms-control treaties that already exist (such as the BTWC and CWC), or have been demanded for a long time (the space-weapons ban). The theoretical advantage that NT as a qualitatively new technology would make preventive limits easier is somehow counteracted here, and the general political problems of achieving meaningful arms limitations become evident.
7 By focusing on the most dangerous NT-enabled applications, implicitly all the others have been accepted, and the limitations proposed cover only a relatively small subset.

7

CONCLUSIONS AND RECOMMENDATIONS

This chapter gives the conclusions of the study. Section 7.1 presents the recommendations for action in the field of arms control proper. Section 7.2 is devoted to transparency and confidence-building measures. Areas to be studied in further research are indicated in Section 7.3, and the final Section 7.4 presents concluding thoughts on international security in an area of revolutionary technological changes.

7.1 Recommendations for preventive-arms-control action for nanotechnology

Based on the considerations of Section 6.4 the actions listed in Table 7.1 are recommended in preventive arms control for nanotechnology (NT).

7.2 Transparency and confidence-building measures

Achieving agreement on formally binding limitation measures may take considerable time, even agreement to start negotiations at all. While such agreements are pending, less formal measures will be very useful – on the one hand to prevent, or at least slow down, dangerous developments within individual states, on the other hand to improve the climate for negotiations on limits.

There is a large potential for mistrust stemming from too little information, in particular in areas where revolutionary changes are foreseen and rapid acceleration of developments is possible (see Section 3.4 for one particular example). As a consequence, the various national NT initiatives – usually acting at a pre-competitive stage – should actively care about mutual transparency. In this respect, the efforts of the US National NT Initiative at international co-operation are to be commended, in particular the inclusion of societal implications (Roco 2001, 2002a). It is positive that the European Commission and the US NSF co-operate in this field (Roco and Tomellini 2002).[1] In the upcoming international dialogue on responsible research and development of NT, military

Table 7.1 Recommendations for preventative-arms-control action

Distributed sensors

- *Complete ban on self-contained sensor systems that are smaller than a certain size limit (3 to 5 cm)*, for the military and civilian sector, starting at the development stage, potentially with exceptions for important civilian applications of smaller sensor systems.
  Verification would rely mainly on on-site inspections with magnifying equipment.

New conventional weapons

- *Uphold the Treaty on Conventional Armed Forces in Europe (CFE Treaty)*, adapt its definitions of treaty-limited equipment if new weapons and carriers with different parameters arrive. Introduce regulation similar to the CFE Treaty in the other regions of the world.
- *Complete ban on small arms, light weapons and munitions that contain no metal*, for the military and civilian sector, starting at the development stage. If weapons and munitions based on non-metallic materials are needed urgently, metallic patterns should be built in obligatorily for a clear signature on x-ray and metal detectors.
  Verification can mainly use on-site inspections with auxiliary equipment.
- *Complete ban on missiles below a certain size limit (0.2–0.5 m)*, for the military and civilian sector, starting at the development stage, with exceptions for innocent civilian applications such as fireworks and life-saving apparatus for stranded ships.
  Verification can rely mainly on on-site inspections with auxiliary equipment.

Body manipulation

- *Moratorium on body implants and other body manipulation that are not directly medically motivated.* The moratorium should comprise the civilian and military sectors, start at the development stage and last for 10 years, with the possibility of prolongation. Concrete interpretations on what constitutes illegal/unethical body manipulation should be developed by international understanding.

Autonomous systems, normal size (above 0.2–0.5 m)

- *General prohibition of re-usable armed, mobile systems without crew.*
- Numerical limits on unarmed, mobile systems without crew, differentiated according to function.
- Prohibition of unarmed, mobile systems without crew in public areas and air space.
- If armed crewless systems cannot be prevented:
  a count them under the CFE Treaty (and potential future similar treaties for other regions);
  b if new systems fall outside of treaty definitions, adapt the latter and agree on numerical limits that strengthen the purpose of the Treaty.
  c *No aiming and weapon release without human decision.*
  d *Prohibition of qualitatively new types of nuclear-weapon carriers.*

  All these prohibitions should apply to the military sector and start at the development stage.
  Verification can rely mainly on on-site inspections.

Table 7.1 Continued

Small mobile systems (size below 0.2–0.5 m)

- *General prohibition of small mobile (partly) artificial systems below a certain size limit (0.2–0.5 m) on/in the ground, on/under water, in the air*, independent of the degree of autonomy and the biological-technical mix, for the military and the civilian sector, starting at the development stage. For important positive applications there should be strictly defined and narrowly limited exceptions:
  a in the civilian sector:
    - for exploration of celestial bodies,
    - for exploration of shattered or dangerous buildings,
    - for inspection of narrow pipes,
    - for surgical operations;

  b in the military sector:
    - for surgical operations.

  For the exceptions, a combination of various technical measures and licensing procedures should prevent abuse.
  Verification can rely mainly on on-site inspections with magnifying equipment.

Small satellites and launchers

- *Comprehensive ban on space weapons of all kinds*, starting at the development stage.
  Verification can to a large part rely on space observation from Earth, augmented by pre-launch inspection after notification.
- Regulation, notification and inspection of small satellites and small launchers.

New chemical or biological weapons

- *Uphold the Chemical Weapons Convention (CWC)*, conclude a clarifying interpretation that for (NT-enabled) agents that are smaller than cells and damage life processes within cells any kind of damaging action counts as 'chemical action' under Article 2.
- *Uphold the Biological and Toxin Weapons Convention (BTWC)*, conclude a clarifying interpretation that (NT-enabled) microscopic systems that can enter the body and are partly or fully artificial are included.
- *Conclude a Protocol on compliance and verification measures* for the BTWC.
- *Conclude a Biosecurity Convention*, with close co-ordination to the BTWC and the new protocol.

In those areas where international agreement may turn out difficult to achieve, or negotiations may proceed only very slowly, export controls should be agreed between the states active in NT research and development (R&D), with the goal to contain proliferation of dangerous NT-based systems and applications.

---

applications should not be overlooked, but actively included, since some of the most problematic risks could come about by military R&D. Codes of conduct for NT should contain rules not only for civilian, but also for military R&D.

Of course, transparency paired with an aggressive military NT programme would not necessarily promote confidence, so some restraint, at least in goals for offence, should accompany transparency.

In parallel to increased transparency, states should negotiate and agree on confidence and security building measures (CSBM), somewhat similar to the ones in force in the Organization for Security and Co-operation in Europe (OSCE) or to the UN Register of Conventional Arms. There is wide scope for potential CSBM in the field of military/dual-use NT R&D: information exchanges on projects and budgets, seminars on strategy and technology, co-operation in projects, exchanges of scientists and engineers.

Formal and informal measures for transparency and confidence-building are important and should be striven for on many levels. Due to the enormous potential of NT, they will probably not suffice to convince the partners that no threatening activities are going on. This can only be the result of legally binding limitation agreements applying to military and civilian R&D that include stringent verification.

7.3 Recommendations for further research

The present work has studied potential military applications of NT from a preventive-arms-control view, necessarily on a rather general level. Future work is needed to deepen the investigations and make recommendations more concrete.

First, military R&D in NT should be followed up continuously:

1 in the areas identified as most problematic here – distributed small sensors, new conventional weapons, body manipulation, large and small autonomous systems, small satellites and launchers, chemical/biological weapons;

- special emphasis should be put on those countries that are active in military high technology but are traditionally less transparent, such as Russia, China, but also France and Israel;

2 for other potential new weapons types, such as electromagnetic acceleration of projectiles, laser weapons or microwave weapons;

3 in the area of nuclear weapons:

- a particular study should be devoted to the possibility of very small pure-fusion weapons.

Subsequently, analyses of preventive limits and verification should be done where appropriate.

Because autonomous combat aircraft are a relatively near-term problem, even without NT, a special project should study them with a view towards preventive arms control.

With a stronger contribution from political science, research should be done on the following topics:

1 Should the international community pursue a new convention, and what should be its topic, e.g. banning hostile use of artificial microscopic systems or manipulation of human physiology?
2 What are the political conditions for acceptance of preventive limitations in NT R&D in various countries – on the one hand, important actors in military high technology, on the other hand exporters of civilian NT that could be used for military purposes? What kind of political process could lead to such acceptance?
3 What would be a good mixture of unilateral restraint, informal agreements, codes of conduct, formal export controls and international preventive-limitation agreements?
4 In case of a nation clearly superior in military NT: would it go to war faster? How would its potential opponents prepare for war?
5 With respect to the dual-use potential of NT R&D, what guidelines should be applied in R&D policy – on the one hand, in R&D funding, on the other hand, in international collaboration?

Finally, again with a stronger natural-science share:

6 Investigate the potential of NT for improved verification of existing arms-limitation treaties (mainly the CWC and the BTWC).
7 Is there a role for NT in the verification of new limits, including the ones proposed here?

Molecular nanotechnology

1 Concerning molecular NT (MNT), the first task is a scientific study of MNT feasibility and the potential time frames of introduction of various applications; it remains to be seen if the studies on molecular self-assembly and on self-replicating nanoscale machines etc., as requested by the US Congress from the National Research Council (Congress 2003), will satisfactorily answer the relevant questions.
2 According to the outcome of that, studies should be devoted to various preventive-arms-control aspects:

- potential military applications of MNT and their dangers,
- interactions between military and civilian R&D,
- appropriate preventive limits in the civilian and military sectors and
- methods as well as technical means of verification in both sectors.

7.4 Concluding thoughts

7.4.1 Preventive arms control for nanotechnology

NT comprises a broad spectrum of different technologies that will have various applications in civilian society and could be used for many different purposes in the military. Evaluation under the criteria of preventive arms control resulted in a relatively small list of most dangerous applications, mostly connected with weapons and utility for offence. Thus, there is a need to limit military NT applications preventively. However, since it is rather complete military systems or missions that create dangers, limitations should focus on these systems or missions, not on the technology used in the implementation. The same recommendation follows from considerations about verification: systems for certain missions can usually be recognized from the outside, without analysing intricate technical details.

NT will in principle allow small and very small systems. Should such systems be introduced into the military, agreement on quantitative limits may become very difficult, on the one hand because they might be produced at very low cost, on the other hand because verification requirements would lead to unacceptable degrees of intrusion. Some indication of this effect can be seen in the US resistance against the compliance and verification protocol to the Biological and Toxin Weapons Convention (see Section 6.4.7). Except for molecules and microorganisms, military systems have not yet shrunk to mini- and micro-size. Thus, the size limits of 0.2–0.5 m for mobile systems recommended here constitute a clear qualitative threshold; if agreement on a prohibition of smaller mobile systems cannot be achieved, then quantitative limits with traditional verification methods may well become impossible.

Because a military ban could not hold for long if, for example, mobile mini- and micro-systems were used widely in civilian society, it is obvious that in many NT areas limits will have to apply more or less equally to both sectors. Fortunately, civilian society will have its own motives to strictly limit such systems – and within democratic societies, where the state has the accepted functions of norm-setting, norm-checking, criminal prosecution and the monopoly of legal force, no great resistance because of security risks or invasion into secrets has to be feared.

7.4.2 Effectiveness and efficiency of military NT applications

While there is no doubt that NT will lead to improved capabilities in many military applications, not all uses that are now in R&D need turn out practical, effective, cost-efficient and sufficiently robust to countermeasures. Deficiencies could stem from many reasons: for small systems, from too little mobility, limited energy supply or payload; for implants and other body manipulation, from human physiology or psychology; for autonomous systems including swarms, from a continuing slow rate of advance in artificial intelligence.

In case of obvious inefficiency, in theory preventive limits would not be needed since the armed forces would not want to introduce such systems.[2] On the other hand, insight about military utility may arrive only at the end; waiting with preventive limits until that time may be too late – in case a system/technology will prove effective despite early expectations, negotiations may have too little time, and military/political motives to introduce the new technology may be much stronger. In fact, one strategy of preventive arms control is to limit systems that seem still remote and have no near-term military utility; in such a case, agreement may be easier – and the limits may turn out militarily meaningful decades later (see Section 5.1.3). Another argument for early negotiations on limits is that the opinions on military utility may differ between countries and governments. There are cases where even obvious inability to fulfil the military task does not dissuade a government from a deployment decision.[3] Of course, in such a case agreement on limitation is impossible and has to wait for changed political conditions.

7.4.3 Military NT needed for international peace operations?

In principle, considerations about international security and peace could lead to the requirement that armed forces carrying out international peace operations or humanitarian interventions – under a UN mandate – should have all kinds of military NT applications available in order to apply force, if needed, as efficiently and selectively as possible. Independent of the general discussion about the justification of such operations, it seems that the narrow scope of preventive limitation recommended here would not much restrain military efficiency in these circumstances.

Of course, very small sensors, new conventional weapons, body manipulation, large and small autonomous systems and small satellites could all add to the fighting capability; one could even stretch the argument and make the point for new chemical or biological weapons.[4] However, there are counter-arguments. One is that the peace force might face similar systems and applications from an opponent who may have got them via proliferation. Second, the marginal increase in the efficiency of peace operations would in no way balance the general dangers of arms race, destabilization, proliferation, terrorist use etc. connected to widespread introduction of such systems.

In many realistic cases for peace operations, there will be a clear technological superiority of the peace forces anyway so that the NT-based applications for which limits and bans are proposed here would contribute only marginally. Thus, a need for peace operations does not put these limits into question.

7.4.4 Central role of the USA

The military policy of the USA – the remaining sole superpower – in general aims at technological superiority and global dominance in armed conflict (e.g. JV 2000). The present administration has strengthened this trend and emphasized unilateral military action, downplaying multilateral agreement including arms limitation.

In particular in military R&D on NT, the USA is outspending the combined rest of the world by a factor of from 4 to 10 (see Section 3.3). As in many other technology areas, also in NT the USA is probably engaged in a virtual arms race with itself. This will likely accelerate the arrival of military systems and applications that should better be prevented. Such applications will then probably find their ways into the hands of potential opponents – states or non-state actors – by many routes: exports, technology transfer, other arms developers following the US example. In turn, these applications would create increasing threats to the USA.

Thus, unilateral restraint by the USA in the most problematic military NT applications could go a long way in preventing such threats or at least delaying their arrival. As long as the technology leader is exploring many possible avenues and sorting out the feasible, it saves others significant R&D investment – of course meaning that they come later. The leader also serves as a role model for all other states active in military high technology. For both reasons, slowing down the advance in military NT by the USA would for a considerable time not result in technological threats from others. Of course, for enduring security from such threats, legally binding multilateral limitation agreements with appropriate verification will be needed. Unilateral restraint by the technology leader could buy sufficient time to work out just these agreements.

This would need a political initiative that one would not expect from the present administration. However, there is also a certain tradition in the USA of restraint and agreed limitation in new military technologies, as demonstrated by the Antiballistic Missile Treaty (1972–2002) and the Laser Blinding Weapons Protocol (1995). The former was conditioned on the Cold War with an ideologically antagonistic superpower as a potential opponent. This situation has changed, but at the fundamental level, military threats still form important parts of national security, not only between the USA, Russia and China.

With NT, revolutionary new military applications are on the horizon that without doubt will be accessible to many actors – Russia and China among them – though with some delay. Metal-free firearms, autonomous micro-robots, selective chemical agents etc. could all lead to very disturbing threats to the security of the USA, posed by military opponents, agents or terrorists. Thus, a hard look at future security in the framework not of narrow military advantage, but of enlightened national interest, could well lead to the insight that international limitations of the most dangerous NT applications would be advantageous for the USA.

As long as the USA is not receptive to such arguments, it may nevertheless be useful for the rest of the international community to prepare, and maybe even conclude, limitation treaties. Even if without the USA other important actors do not become parties, such agreements would create obligations for relevant countries and will have some political effect inside the USA.[5]

7.4.5 Further development of the international system?

NT together with the other 'converging' technologies that it will enable – biotechnology with genetic engineering and manipulation of cell mechanisms, information technology with ubiquitous small computers, new levels of artificial intelligence and large and small robots etc. – will create not only ethical challenges, but many possibilities for misuse. Pervasive and often very small, these technologies will be difficult to control. Within civil societies, the general mechanisms of democratic decision, legislation, licensing, inspection and criminal prosecution may be sufficient to guarantee security for the members and institutions of society.

In order to ensure – and sometimes enforce – compliance with the rules, civilian societies have come to accept some relatively strict limits and fairly wide-ranging rights for inspection by state personnel – e.g. in the areas of work safety, environment, correct accounting and criminal investigation. For small and pervasive systems, similar degrees of inspection would be required on the international level to verify compliance with agreed limits. This is conceivable in the civilian sector – but even here the fear of invading into commercial secrets will create problems. In the military sector, on the other hand, limitations as such as well as intense inspection will meet strong resistance, since exploitation of new technology as well as secrecy are essential parts of preparations for armed conflict.

As a consequence, one arrives at the problem that the traditional way of guaranteeing national security – namely by the threat of armed force – may be no longer compatible with the advance of technology. If the security threats following from new technology can only be contained by intense inspections that would endanger the very functioning of the armed forces in their central role, security can no longer be reliably ensured by national armed forces.

In this vein, the powerful new technologies that require stringent international control could act as a catalyst in a process of moving away from the security dilemma and strengthening civilian-society elements in the international system. This would mean strengthened international institutions and international law, in particular criminal law with prosecution of perpetrators, moving into a direction towards an international monopoly of legitimate force, strong enough to prevent or punish threats or use of illegal force. In such a process, reliance on national armed forces for security could be reduced step by step.

APPENDIX 1

GENERAL NANOTECHNOLOGY LITERATURE

Due to the vast amount of literature in nanotechnology, only a few handbooks, a selection of scientific journals and a few Internet sites are given here.

Handbooks

Bhushan B. (ed.) (2004) *Springer Handbook of Nanotechnology*, Berlin etc.: Springer.
Goddard III W.A., Brenner D.W., Lyshevski S.E., Iafrate G.J. (eds) (2002) *Handbook of Nanoscience, Engineering, and Technology*, Boca Raton FL: CRC Press.
Nalwa H.S. (ed.) (1999) *Handbook of Nanostructured Materials and Nanotechnology*, 5 volumes, San Diego etc.: Academic Press.

Selected scientific journals

Applied Physics Letters (section Nanoscale Science and Design)
Fullerenes, Nanotubes, and Carbon Nanostructures
IEEE Transactions on Nanotechnology
Journal of Nanoparticle Research
Journal of Nanoscience and Nanotechnology
Journal of Vacuum Science & Technology B: Microelectronics and Nanometer Structures
Materials Letters
Nanostructured Materials
Nano Letters
Nanotechnology
Physica E: Low-dimensional Systems and Nanostructures
Precision Engineering: Journal of the International Societies for Precision Engineering and Nanotechnology
Virtual Journal of Nanoscale Science and Technology. Online, available at: ojps.aip.org/journals/doc/VIRT01-home

APPENDIX 1

Some Internet sites

www.nanoforum.org
nano.gov
www.smalltimes.com
www.foresight.org

APPENDIX 2

US DARPA NT-RELATED EFFORTS

Table A1 DARPA programs that are (potentially) related to NT in a narrow and a broader sense. The latter includes aspects of biology, artificial intelligence, cognitive science and robotics. Programs that deal mainly with software, database handling, surveillance for homeland security, information warfare etc. as well as pure MST projects are not included. The list is based on a perusal of the budget estimates (DARPA Budget 2003); for time reasons, no additional information was sought. Thus, some NT-related projects may be missing because this was not identified in the description. Some projects (such as for large or small robots) may not yet include NT, but are likely to do so soon at least in the form of smaller, more capable computers. The planned expenses for FY 2003 are given in US$ million (rounded to one million); in several cases, only a (small) part will be NT-related (e.g. with small-satellites or materials); partial expenses are not available. Projects that ended before 2003 are excluded; projects that will start later are given with zero funding. Expenses for programs narrowly related to NT are given in bold; those broadly related to NT are given in medium typeface

Budget Activity/ *Program Element/*Project	*Program*	*Remark*	*FY 03*
BA1 Basic Research			
Defence Research Sciences			
Bio/Info/Micro Sciences BLS-01	BioComputational Systems	computing mechanisms in the bio-substrate, miniaturized hardware, models and software tools for cellular processes	**30**
	Simulation of Bio-Molecular Microsystems	molecular recognition, signal transduction, micro- and nano-scale transport; biological-synthetic integration	**15**
	Bio Futures	integration of information technology and biological processes	**10**

continued

Budget Activity/ *Program Element*/Project	*Program*	*Remark*	*FY 03*
	Biological Adaption, Assembly and Manufacture	investigate and exploit adaptation to harsh environments	**10**
	Nanostructure in Biology	understand and exploit biological materials	**9**
	Brain Machine Interface	access neural codes in brain, integrate into peripheral device/system	12
Information Sciences CCS-02	Computer Exploitation and Human Collaboration	intuitive interaction with computers	24
Electronic Sciences ES-01	Supermolecular Photonics Engineering	engineer molecular structure for optical properties	**0**
Materials Sciences MS-01	Nanoscale/Bio-molecular and Metamaterials	for electric drive, power electronics, imaging	**13**
	Spin Dependent Materials and Devices	light-emitting diode, transistor, quantum logic	**19**
	Engineered Bio-Molecular Nano-Devices and Systems	real time observation and analysis of bio-molecular signals	**7**
	Spin Electronics	organic electronics, nanostructures	**15**
	Ultra Performance Nanotechnology Center		**3**
	Joint Collaboration on Nanotechnology		**2**
	Center for Nanostructure Materials		**0**
	Nanotechnology Research and Training Facility		**2**
	Molecular Electronics		**1**

BA2 Applied Research			
Computing Systems and Communication Technology			
Intelligent Systems and Software ST-11	Software for Situational Analysis	large knowledge bases	18
	Taskable Agent Software Kit (TASK)	multi-agent systems for a global objective, e.g. autonomous vehicles	11
High Performance and Global Scale Systems ST-19	Autonomous Systems Control	mission planning, mobile robots	6
	Mixed Initiative Control of Automa-Teams	planning, assessment and control of distributed, autonomous combat forces	17
Language Translation ST-29	Situation Presentation and Interaction	face-to-face speech translation in foreign territories, automatic dialogue for C2	9
	Automated Speech and Text Exploitation in Multiple Languages	access foreign speech and text, create transcripts, detect critical intelligence	34
Cognitive Systems Learning and Perception ST-30	Perceptive Assistant that Learns	cognitive systems using past experience and knowledge, do purposeful perception	7
	Real-World Learning Technology	cognitive systems improving their performance and understanding over time	0
Communications, Interaction and Cognitive Networks ST-31	Augmented Cognition	measure and manipulate a subject's cognitive state	19
	Collaborative Cognition	collaborative agents in dynamic multi-agent environments	2
	Self-Aware Peer-to-Peer Networks	resilient, scalable sensor-computation networks with decentralized control	2
Cognitive Systems Foundations ST-32	Network-Centric Infrastructure for Command, Control and Intelligence	virtual work centres bringing together a combination of people, computer systems, robots, and data	9
	Architectures for Cognitive Information Processing	perception, reasoning and representation, learning, communication and interaction	0
Knowledge Representation and Reasoning ST-33	Autonomous Software for Intelligent Control	autonomous mobile robots, near-human performance in autonomous vehicle navigation and interaction with humans	15
	Knowledge Based Systems	from large, strategic knowledge banks to personal knowledge pads	0
	Advisable Systems	user control in natural and flexible ways	4

continued

Budget Activity/ *Program Element*/Project	*Program*	*Remark*	*FY 03*
Embedded Software and Pervasive Computing			
Software for Autonomous Systems AE-02	Common Software for Autonomous Robotics	software technologies for large groups of resource-constrained micro-robots	5
	Software Enabled Control	for unmanned and manned aircraft	18
Software for Embedded Systems AE-03	Large Scale Networks of Sensors	software for distributed micro-sensor networks	4
Biological Warfare Defence			
Biological Warfare Defence Program BW-01	Advanced Diagnostics Sensors	rapid detection of pathogens	5
		biochips etc. for BW agents	**37**
Tactical Technology			
Advanced Land Systems Technology TT-04	Collaborative Munitions	distributed sensing, communication, understanding environment	6
	Close-In Sensing/Odortype Detection/Dynamic Optical Tags	detectors for individual chemosignals, small retroreflecting tags	13
Advanced Tactical Technology TT-06	Varuna	ultra miniature audio and video recording, tracking and locating systems	0
Aeronautics Technology TT-07	Long Endurance Hydrogen Powered Unmanned Air Vehicle	intelligence, surveillance and communications equivalent to low satellites	4
	Organic Air Vehicles (OAV) in the Trees	micro-air vehicle flight under canopy and in buildings	4
Network Centric Enabling Technology TT-13	RoboScout	low-cost, small, mobile sensors for close-in reconnaissance	0
	Eyes-On System	micro-air vehicle to enter and survey a target area	0
	Urban Robotic Surveillance	mobile sensor systems including ground and air platforms	0

Materials and Electronics Technology			
Materials Processing Technology MPT-01	Structural Materials and Devices	light-weight, multifunctional materials including processing for all media	**33**
	Smart Materials and Actuators	smart materials and actuators, self-diagnosis and -repair, increase soldier's physical capabilities	**31**
	Functional Materials and Devices	magnetic sensors, magnetic memory, conductive/ electroactive polymers, direct-write conformal electronic	**28**
Microelectronic Device Technologies MPT-02	Chip Scale Atomic Clock	atomic-resonance-based time-reference	**13**
	Technology Efficient, Agile Mixed Signal Microsystem	CMOS integrating >10,000 nanoscale devices	**12**
	Interfacing Nanoelectronics	new interconnects	**0**
	Nanostructured Photonic and Biomedical Materials	self assembly for spintronics and bio-inspired optics	**1**
Beyond Silicon MPT-08	Quantum Information Science and Technology	decoherence, signal attenuation, algorithms and protocols, scalability	21
	Moletronics	scalable devices from molecules, nanotubes, nano-wires etc., hierarchical self-assembly	**23**
	Molecular Computing	integrate memory with control logic and data paths	**0**
Biologically Based Materials and Devices MPT-09	Bioinspired and Bioderived Materials	actuators, adhesives, sensors, locomotion, optics, biomolecular motors, animal sentinels	**35**
	Biochemical Materials	preserve tissue, cells; overcome sleep deprivation; enhance endurance, mimimize caloric intake	38
	Responsive Membrane Devices	microreactors, nanotechnology; single-molecule DNA/RNA/protein sequencing	**7**
	Bio-Magnetic Interfacing Concepts	nanomagnetics for detecting, manipulating and controlling in single cells and biomolecules	**5**
	Energy from Bio Systems	liquid-fuel, biocatalytic fuel cell	0

continued

Budget Activity/ *Program Element*/Project	*Program*	*Remark*	*FY 03*
BA3 Advanced Technology Development			
Advanced Aerospace Systems			
Advanced Aerospace Systems ASP-01	Advanced Air Vehicle: A160 Hummingbird Warrior	vertical take-off and landing unmanned air vehicle	13
	Unmanned Combat Air Vehicle	for suppression of enemy air defence and strike missions	59
	Naval Unmanned Combat Air Vehicle	for suppression of enemy air defence/strike/ surveillance missions	22
	Joint Unmanned Air Vehicle	for suppression of enemy air defence/strike/ surveillance missions	0
	Unmanned Combat Armed Rotorcraft	for armed reconnaissance and attack missions	23
Space Programs and Technology ASP-02	Orbital Express Space Operations Architecture	robotic, autonomous on-orbit refuelling and reconfiguration of satellites; small satellite	**40**
	Responsive Access, Small Cargo, Affordable Launch	low cost orbital insertion of small satellite payloads	24
	Rapid On-Orbit Anomaly Surveillance and Tracking	small-satellite constellation to detect and track on-orbit objects	0
	Space Assembly and Manufacture	including robotic assembly and small satellites	0
Advanced Electronics Technology			
Advanced Lithography MT-10	Advanced Lithography	complex microelectronics patterns at sub 0.05 μm resolution	**24**
	Laser Plasma X-Ray Source		**3**
	Advanced Lithography Demonstration	point source lithography	**4**
	Advanced Lithography X-Ray Thin Film Development		**4**

MEMS and Integrated Micro-systems	Bio-Fluidic Chips (BioFlips)	assessment of the warfighter's body fluids with therapy	13
Technology MT-12	MEMS Mechanical Computation and Data Storage	data as tiny pits, phase changes, molecular changes	0
Mixed Technology Integration MT-15	Nano Mechanical Array Signal Processor	nano mechanical structures for radio frequency signal processing	**17**
	Digital Control of Analog Circuits RF Front Ends	nano-CMOS, microfluidics, electronics, MST	**16**
	Flexible Nanocomposite Organic Photovoltaic Cells	200× increase in power/weight	**0**
Command, Control and Communication Systems			
Command and Control Information Systems CCC-01	Advanced Ground Tactical Battle Manager	expands operational plans for robotic forces into commands for each tactical vehicle	4
	Banshee	data links expendable for platforms (weapons, small unmanned air vehicles)	0
	Comprehensive Force Protection	sensors and mobile platforms for surveillance around camp	0
Sensor and Guidance Technology	–		–
Marine Technology	–		–
Land Warfare Technology			
Rapid Strike Force Technology LNW-01	Tactical Mobile Robotics	for missions in complex environments	1
	Micro Air Vehicle Advanced Concept Technology Demonstration	backpackable reconnaissance and surveillance system	4
Future Combat Systems LNW-03	FCS Supporting Technologies	perception for robotics, unmanned ground combat vehicle, micro air vehicles, integrated command and control	77

continued

Budget Activity/ *Program Element*/Project	*Program*	*Remark*	*FY 03*
Network-Centric Warfare Technology			
Joint Warfare Systems NET-01	Confirmatory Hunter Killer System	loitering weapon/unmanned air vehicle	0
Maritime Systems NET-02	Piranha	including swarms of mini unmanned underwater vehicles for sensing sea targets	0
Total NT-related narrow	**36**		**465**
broad	51		547
	Programs		*Exp.*

NOTES

1 INTRODUCTION

1 NNI 2002: 11.
2 Shorter presentations have been given in Altmann and Gubrud (2002), Altmann (2004).
3 The Foresight Institute was founded in 1986 by K. Eric Drexler, Christine Peterson and James Bennett (Foresight 2003a).
4 The term 'nanotechnology', used by Drexler beside 'molecular engineering', had already been coined in a different sense in 1974 by N. Taniguchi (Japan) (Franks 1987).
5 A review of NT to the mid-1980s is given by Franks (1987); a detailed overview of NT-related developments to the late 1980s is provided by Schneiker (1989). For short history lists or tables see, e.g. Crandall 1996a; Smith II 1998: Appendix A; Scientific 2001: 36; see also Drexler 1992: App. B.
6 The ETC (Erosion, Technology, Concentration, formerly RAFI) Group is an international civic-society organization based in Canada.
7 The report was debated in November 2003 in the German Parliament.
8 In his afterword to the 1990 reprint, Drexler (1986/1990: 241) stated that there would be little incentive to build replicators capable of surviving in nature (see also Merkle 1992). A recent quantitative analysis argues that the risk is limited and can be contained (Freitas 2000).
9 In Germany, the national newspaper *Frankfurter Allgemeine Zeitung* reprinted Joy's article and carried a series of articles relating to it in summer 2000.
10 The workshop was part of the preparation of the Institute for Soldier Nanotechnologies (see Section 3.1.6).
11 The 'two-weeks revolution' has later been called improbable, see note 15 in Chapter 2.
12 0) slow growth of technology, resource conflicts; 1) fast acceptance and arrival of assemblers, then general death from a military-built replicator; 2) following environmentalists and arms-control activists, the public suppresses MNT, but then destruction starts from an uncontrolled place; 3) nations develop MNT in technological rivalry, including a multilateral arms race; 4) development of MNT mainly in international co-operation with the industrialized democracies leading. Note that some of the scenarios are denoted as absurd.
13 A short article had been published previously (Gubrud 1989).
14 The Institute's website is http://www.foresight.org; see note 3.
15 'The idea of a killing system without direct human control is frightening. Because of this, developing the rules of engagement for robotic warfare is likely to be extraordinarily contentious.'

16 More general statements were made by the Joint Doctrine and Concepts Centre (JDCC 2003).
17 Most parts of the discussion are centred around existing military R&D, see Section 3.1 of the present work.
18 The article does not clearly differentiate between MST and NT; questionable are, e.g. the statements that MST and NT were developed decades ago for use in nuclear artillery shells, and that insensitive high explosive is isolated from small igniter charges that, on arming, are moved into position by micro-electromechanical systems.
19 The NIF, situated in the Lawrence Livermore National Laboratory, measures 200 m by 85 m; the light pulses from 192 laser-amplifier chains are focused from all directions on a mm-size fusion capsule in the centre of a 10-m-diameter sphere (Heller 1998, 1999; Parker 2000).

2 OVERVIEW OF NANOTECHNOLOGY

1 For general NT overviews, see Taniguchi 1996; Timp 1999; Nalwa 1999; Gross 1999; Scientific 2001; Goddard III *et al.* 2002; Bhushan 2004.
2 The US NNI uses a lower limit of 1 nm (NNI 2000: 19; NNI 2002: 11).
3 For definitions of NT which stress the novelty aspect, see also IWGN 1999: vii; NSET 2000; NNI 2000: 19f.; NNI 2002: 11.
4 In English, MST often comes under the heading of micro-electromechanical systems (MEMS) which, however, is narrower; sometimes, NEMS is used for the nanoscale version.
5 However, in solid-state physics 'mesoscopic' is also used for systems of sizes around 30 nm, the cross-over regime where many quantum-mechanical phenomena due to electron confinement and coherence become observable.
6 For general applications including fabrication of smaller systems, see, e.g. MacDonald 1999; for the 'Millipede' project of writing and reading data by an array of heated cantilever probes that indent a polymer film, see Vettiger *et al.* 2002; IBM 2002.
7 The brochure predates the NNI, but is part of its material for the public.
8 As a general size indicator, the dynamic-random-access-memory (DRAM) half pitch (half spacing between the densest parallel (metal or polysilicon) interconnects) is used; the roadmap contains extrapolations for many other characteristics (SIA 2001; ITRS 2002).
9 A quantum dot is a nm-sized structure that can contain a single electric charge. Roughly similar to the conditions in an atom, the charge can exist in various states with different energy levels. Jumps between such states can occur with absorption or emission of a light quantum carrying the energy difference.
10 Ballistic transport means that electrons move at high speed without being scattered at impurities.
11 Of course, some branches of nanoscience and NT are already now working on/with molecules. To differentiate the visionary concept, a term such as 'assembler-based NT' would be more exact. However, for consistency with established practice, I use the designation MNT.
12 A very comprehensive review of the history of NT from the 1950s to the late 1980s with many references was given by Schneiker (1989). Concerning the works by Drexler (1986, 1981), Schneiker remarks (p. 464) that 'references to the real originators of many of the ideas that he [Drexler] discusses are not given'.
13 These have later been called Feynman machines, see Schneiker 1989. The invention of the scanning tunnelling and atomic-force microscope provided direct ways of manipulating single atoms.

14 The program could be carried in the assembler, e.g. in the form of a linear molecule similar to DNA; in this case the system could be self-reliant and potentially proliferate in an uncontrolled way. To prevent that, the program could be broadcast every time, e.g. acoustically, see Drexler 1992: Section 16.3.2, Merkle 1999.
15 Kaehler (1996) dates the 'two-weeks revolution' idea to 1980.
16 V. Vinge has argued that with growth in computing power and software capabilities to human and superhuman capabilities change would take place at ever increasing speed, leading to a 'singularity' (expected between 2005 and 2030) beyond which prediction will be impossible: Vinge 1993 and refs, see also Hanson 1998.
17 'Transhumanists everywhere ... must join to break the chains that bind science' (Bainbridge 2003).
18 Kurzweil's (1999) fictitious future person 'Molly' in 2099 alleges that 'Drexler has written a series of papers showing the feasibility of building technology on the femtometer scale, basically exploiting fine structures within quarks to do computing' (p. 244). This is ascribed to a future Drexler continuing as an enhanced/cyber entity – in flat contradiction to the real Drexler's arguments of 1986. Kurzweil makes the allegation his own in the corresponding note (p. 342) where from extrapolating Moore's law he predicts 'picoengineering' in 2072 and 'engineering at the femtometer' in 2112.
19 Drexler (1986: Ch. 10) makes a similar argument about engineering in a collapsed star. Note that Moravec (1988: 74) speculates that ultradense matter as in collapsed white dwarfs and neutron stars might some day be exploited. For physics estimates of the theoretical limits on computing by extremely hot and dense matter and of the universe at large, see Lloyd 2000, 2002.
20 The request of a RAND report (Nelson and Shipbaugh 1995) for 'a detailed and objective technology assessment examining the current status and likely prospects of molecular technology' has not yet been fulfilled. That report identified several intermediate steps that can serve as indicators of molecular NT approaching or as terminal points.
21 Several such articles have been published in the scientific journal *Nanotechnology*, based on contributions from the Foresight Conferences on Molecular Nanotechnology, e.g. Merkle 1997, 1999, 2000; Hall J.S. 1999. On medical nano-robots, see for example Haberzettl 2002.
22 I got a similar impression in several discussions with NT scientists.
23 Recent Feynman-Prize winners are (experimental / theoretical): 1999: P. Avouris (IBM) / W.A. Goddard III, T. Cagin, Y. Qui (Caltech); 2000: R.S. Williams (HP), P. Kuekes (HP), J. Heath (Univ. of California Los Angeles) / U. Landman (Georgia Tech); 2001: C.M. Lieber (Harvard) / M.A. Ratner (Northwestern Univ.); 2002: C. Mirkin (Northwestern Univ.) / D. Brenner (North Carolina State Univ.) (Foresight 2003).
24 C. Montemagno, then associate professor at the Dept. of Biological Engineering, Cornell University, USA, used the title 'Nanomachines: A roadmap for realizing the Vision'; the research is described in Soong *et al.* 2000.
25 With different degrees of views on transcending humans, e.g. Moravec 1988; Kurzweil 1999; Brooks 2002. For NT computing experts who treat sentient artificial intelligence as a realistic possibility, see, e.g., Williams and Kuekes 2001; Williams 2002. For the general discussion on 'strong AI', see Russell and Norvig 2003: Ch. 26.
26 For several years, the Foresight Conferences on MNT (http://www.foresight.org/Conferences) have reduced the discussion of assemblers and focused more on actual NT research. The US NNI seems to downplay links to visionary NT;

however, recently it has embraced several of its concepts, see Section 2.3. See also note 17.

27 NSET is a subcommittee of the US Committee on Technology of the National Science and Technology Council; it co-ordinates the Federal multi-agency nanoscale R&D programmes, including the NNI. Its chair is Mihail C. Roco (NSF), the Executive Secretary is James N. Murday (NRL) (NNI 2002: 2, 55).

28 For the military part see Section 1.5.10. A successor conference was held in February 2003, see Infocastinc 2003.

29 Among the eleven sponsoring agencies, six were military: AFOSR, ONR, ARO, ARL, DARPA, BMDO. The others were: NSF, DoC (incl. NIST and Technology Administration), NIH, NASA (Siegel *et al.* 1999: xvii, 1).

30 For the history leading to the NNI see Smith II 1998.

31 'What we are seeing is a global race to keep up with nanotechnology research and development' (Harper 2002).

32 Note that this refers only to the EU funding and that the figures for Europe in Table 2.7 include national expenses.

3 MILITARY EFFORTS FOR NANOTECHNOLOGY

1 The Strategic Research Objectives of the Basic Research Plan were: biomimetics, nanoscience, smart structures, broadband communications, intelligent systems and compact power sources, e.g. Killion 1997.

2 DoD ($32 million, sum of DARPA, ARO, ONR, AFOSR) was second to NSF ($65m), the other agencies were DoE ($7m), NIH ($5m), NIST ($4m) and NASA ($3m).

3 Two programme elements (BLACK LIGHT and small business) were finished in 2002 and are not listed in Table 3.4.

4 In this program, contracts have been let to two German institutions: Fraunhofer Institut für Autonome Intelligente Systems (Scorpion, ambulatory robot) and Universität Bonn, Institut für Zoologie (insect infrared sensors) (DARPA CBS 2003 and links).

5 For functional magnetic-resonance imaging to develop means to augment human cognition, in particular when doing complex tasks in interaction with computers, see DARPA AugCog 2003.

6 This multidisciplinary institute is different from the traditional NRL structure. A new Nanosciences Building was to become available in 2003 (NRL Nanoscience 2003).

7 Interagency Working Group on Nanoscience, Engineering and Technology, IWGN 1999; Subcommittee on Nanoscale Science, Engineering and Technology (NSET) (NNI 2000, 2002).

8 The other four are at non-weapons laboratories (Oak Ridge, Berkeley, Brookhaven, Argonne) (DoE 2003).

9 The other focus is Materials Under Extreme Conditions (MRI 2003; CMS 2002: 36).

10 Nanotechnology for the Soldier System Conference and Workshop, 7–9 July 1998, sponsors Natick Research, Development and Engineering Center of Army Soldier Systems Command, Edgwood Research, Development and Engineering Center of Army Chemical and Biological Defense Command, ARO, NSF, ARL (Natick 1998).

11 Mullins (2002) reports an industry contribution of $15.3 million and one from MIT of $14.7 million; the Army investment is planned to continue for ten years or more.

12 Bundesministerium der Verteidigung, Rü IV, letter of 30 Jan. 2003.

13 According to INT members (Dec. 2003), the study will be published.
14 http://www.qinetiq.com. This firm was founded together with the MoD Defence Science and Technology Laboratory (http://www.dstl.gov.uk) in July 2001 when the former Defence Evaluation and Research Agency (DERA) was dissolved.
15 Within the Applied Vehicle Technology Panel (AVT) of the NATO Research and Technology Organization (RTO).
16 See http://www.cordis.lu/nanotechnology/src/intlcoop-ru.htm.
17 I thank Mark Gubrud for pointing out this misrepresentation. In Altmann (2001: 94–95), I had taken Pillsbury's statement at face value.

4 POTENTIAL MILITARY APPLICATIONS OF NANOTECHNOLOGY

1 As mentioned in Section 1.2, many examples have only appeared in the course of this project.
2 A similar size applies to circuits passively energized over a few m distance by radio frequency/induction coil.
3 Derived from 10^{11} neurones in the brain with on average 10^3 synaptic connections each, doing 200 pulses per second (see Kurzweil 1999: Ch. 6); energy considerations lead to similar values (Merkle 1989).
4 For a futuristic civilian conception of such a vehicle, see BMBF 1998: 6–7.
5 Glasstone and Dolan (1977: 13) wrote that because of experimental uncertainty (3.8–4.8 MJ/kg), the value of 10^{12} cal/kg = 4.2 MJ/kg was defined for the usual measure of energy yield of nuclear and conventional bombs in units of kilotons (= 10^6 kg) TNT. Ullmanns *Encyklopädie* (1982: Table 5) gives 4.52 MJ/kg, Kubota (2002: Table 4.3) lists 5.07 MJ/kg.
6 For the purpose of penetration, the isotopic composition of uranium is not important. *Depleted* uranium (with about 0.2 per cent uranium-235 as opposed to 0.7 per cent in natural uranium) is being used because in the USA there are hundreds of thousands of tons stockpiled as tails from the enrichment of uranium-235 for nuclear weapons (to above 90 per cent) and for reactor fuel (to about 3 per cent).
7 Bulk amorphous metals or metallic glasses are alloys, often of five different metals, which do not form crystallites because the atoms are chemically different (size etc.). They can support about double the elastic strain of ordinary multi-crystalline materials. Such material (Vitreloy 1, containing zirconium, titanium, copper, nickel and beryllium) is already being used commercially in golf clubs. See Johnson (1999) who already mentioned its utility for penetrators.
8 Magness *et al.* (2001) used an alloy of hafnium, titanium, nickel, copper and aluminium with 1.1 Mg/m^3 density, free of the toxic beryllium. See also Dowding 2003 and Army RDT&E 2001. For the DARPA program 'Structural Amorphous Metals' which goes much beyond penetrators, see Section 4.1.4.
9 Magness *et al.* 2001 write from a US perspective. Note that other countries such as Germany and Switzerland banned depleted uranium and are using tungsten alloy, Lanz *et al.* 2001.
10 NRL projects deal with nano-channel glass arrays that could connect to thousands or millions of neurones, and two-dimensional multiplexer arrays for them (Kafafi 2003: 18–22).
11 The US Predator aircraft was equipped with a Hellfire missile; this was used to kill six putative terrorists under remote control in Yemen in November 2002 (Hoyle and Koch 2002).

12 For example, the European Fighter Aircraft is designed for 9 g vertical acceleration, but due to the pilot the flight is limited to 7 g (1 g = gravity acceleration at sea level = 9.8 m/s^2) (Altmann 2000: 202).
13 For more information on MST-based mini-/micro-robots and earlier military R&D projects, see Altmann 2001: Section 4.2.15.
14 There is a grey area to intelligent small missiles and target-seeking small munitions.
15 The first generation of nuclear weapons used pure fission (uranium or plutonium), the second added fusion (boosted fission, two-stage (hydrogen) bomb), and the third enhanced special effects, such as neutron radiation or the electromagnetic pulse.
16 See note 19 in Ch. 1.
17 The CWC concerns only chemicals against humans or animals, see Section 6.4.7. However, widespread use of herbicides modifying the environment is outlawed by the Environmental Modification Convention of 1977 (ENMOD 1977).
18 However, secret military research on incapacitating agents including opiates (not related to NT) has been carried out in the context of work on non-lethal weapons (Sunshine 2004). On the dangers to the CWC and BTWC from non-lethal-weapons work see CBWCB 2003.
19 Artificial systems could put the applicability of the BTWC into question. On the blurring of the dividing line between chemical and biological weapons and a discussion of a potential new Convention see Section 6.4.7.
20 However, partly secret military research on genetically engineered microbes that would degrade material or act as taggants (not related to NT) has been carried out (Sunshine 2002, 2002a). Development for hostile uses is prohibited by the BTWC.
21 The only exception would exist if micro-nuclear weapons became at all practical and could be mass-produced cheaply.
22 Note that unfortunately the terms micro-, nano-, pico-, femto-satellite have been used to designate masses below 100 kg to 0.1 kg, with steps of a factor 1/10 (e.g. SSHP 2001). This is inconsistent with the general use of these prefixes in SI units (factor 1/1000) and should be revised.

5 PREVENTIVE ARMS CONTROL: CONCEPT AND DESIGN

1 Producer liability for damage adds another layer of protection. Note, however, that most of such rules have only been introduced after extreme damage occurred, and sometimes only very late.
2 There is of course great cultural variety here; take, for example, beyond R&D, the case of firearms which are banned in most civilized societies, but in the USA are seen by a majority as an important part of private protection and deterrence against other people's arms.
3 Defensive superiority would mean a reduced capability to deploy forces at long range and to use long-range weapons. Reconciling this with a need for global crisis intervention needs creativity; quantitative and qualitative limits as well as systematic involvement of the UNO could help.
4 This is rarely discussed in public, but nevertheless is likely to be part of fundamental convictions. One case when such considerations were brought to the public was the Draft Defense Planning Guidance prepared under P.D. Wolfowitz, then Under Secretary for Policy in the US DoD: Nuclear proliferation could motivate Germany, Japan and others to acquire nuclear weapons them-

selves, leading to global competition with the USA and, in a crisis, to military rivalry (Tyler 1992). Later, the text was changed and not published.

5 While such arguments are often made, e.g. in the debate on allegedly insufficient technological efforts of the European members of NATO versus the USA, it is remarkable that there is no discussion that the USA should decelerate its rate of innovation in the interest of better co-operation.

6 The problem of mistrust of the user countries in the motives of the supplier countries needs to be solved, optimally by a regime including the former.

7 In particular, there have been differences of opinion about the inclusion of the arms-race topic in II. Because of the conceptual differences between dangers to humans and to environment/sustainability in III, both topics are treated separately here, differently from Altmann (2001).

8 For a ban on laser weapons (for damaging material, not eyes or sensors), military lasers with outside beam propagation would be banned above 100 watts average power and 0.15 m beam diameter. For civilian high-power lasers with outside beam propagation, three threshold levels on the laser brightness and two on mirror diameter were proposed with inspection rights and increasing levels of notification, licensing and special precautions against military misuse (Altmann 1986, 1994).

9 The Ad Hoc Group was mandated in 1994; before, meetings on verification of government experts had taken place 1992–1993.

10 On the precautionary principle see Section 5.1.1. Concepts for application of the precautionary principle to international security questions still need to be developed. Here one finds an interesting contradiction in the traditional approaches: whereas in the civilian sector unambiguous proof of damage was demanded before new technologies could be limited, in the military field efforts to strengthen one's armed forces were often based on not proved worst-case assumptions.

11 Note that the CTBTO exists as a Preparatory Commission since the Treaty has not entered into force.

12 The nuclear-weapons states can of course use the experience of their previous tests. With immensely faster computers, three-dimensional full modelling of nuclear explosions may enable R&D even of new warhead types (see Section 4.1.19.2), even though one can doubt whether the military will introduce new weapons that have not been tested in reality.

13 A more differentiated view would state that innovation is a process with many sub-cycles. During testing of a new system, for example, a problem may show up which requires that a certain component goes back to the research stage. Development goes on after first deployment and systems are routinely upgraded.

14 A more complete picture would include stockpiling, transfer, assistance. These and later stages of the life cycle of a military technology (upgrades, taking out of service – with destruction or export to less-developed countries) are not relevant for preventive arms control.

15 The US Pharmaceutical Research and Manufacturers of America have since the mid-1990s raised strong objections to some forms of non-challenge visits in laboratories, arguing that these would place proprietary information at risk. The US government has taken up this argument and left the negotiations in 2001; another motive was probably the protection of secret controversial military activities. See Nixdorff *et al.* 2003: Section 8.6; Rissanen 2002; Feakes and Littlewood 2002.

16 A laser focus on the order of 10 m could start fires or melt metal; with microwaves, the size would be kilometres with a power density not much above that of sunlight (Altmann 1986: Section 7.3.5.9, 1994).

6 PREVENTIVE ARMS CONTROL CONSIDERATIONS FOR NANOTECHNOLOGY

1 Art. 2(2) of the Anti-Personnel Mine Convention: '"Mine" means a munition to be placed under, on or near the ground or other surface area and to be exploded by the presence, proximity or contact of a person or a vehicle' (APMC 1997).

2 In the case of laser blinding weapons, for example, the USA had argued until early 1995 that blinding is no more cruel than killing; later that year, however, it signed the Protocol banning laser blinding weapons (Morton 1998).

3 Pre-emptive attack to negate an immediately impending attack by the opponent is different from *preventive* attack that aims to reduce the military power that an opponent may bring to bear at some time in the future. While arguments can be made that the first type can fall under legitimate self-defence under Art. 51 of the UN Charter, the second is clearly outlawed; otherwise, nearly any war could be justified.

4 This qualification was missing in Altmann (2001: Ch. 6).

5 Western export controls prohibited exports of the respective recent microprocessor generation to the countries of the former WTO even though millions of them were used within the civilian societies.

6 One might argue that medical soldier systems that reduce the effects from battle injuries would reduce dangers to humans, but generalizing this strand of argument would fast lead to inconsistencies, e.g. requests for stronger armour and more effective offence. Thus, human damage afflicted in armed conflict has to be treated in a separate framework.

7 This has also been mentioned by Brendley and Steeb (1993: 30).

8 Of course, smaller size and mass (on the order of 10 cm and 1 kg) would make stealing and covert transport easier, but existing nuclear weapons of 1 m and 100 kg pose no special hurdles either.

9 Already now people using cellular phones make their actual location known to within a few kilometres (they link to the closest relay station). With future high-bandwidth connections, relay stations may be every 100 m or so, in buildings much closer.

10 On the arguments and lawsuits about forced anthrax and multiple vaccinations in the US armed forces, see Sunshine 2004a.

11 Note that some MST applications that could be advanced by adding or using NT have not been included here. This concerns the categories optics, fluidics, inertial measurement/guidance/stabilization, safety/arming/fusing, identification friend/foe, remote detection of chemical substances, micro chemical analysis, biotechnological/biomedical analysis *in vitro*, implanted location/identification devices and mini-/micro-mines. For their evaluation see Altmann 2001: Ch. 6.

12 The non-ratification of the Comprehensive Test Ban Treaty by the USA and the intention of the present administration to shorten the time for a resumption of nuclear tests show that the military motive continues to exist, but demonstrate also the danger to the Treaty or the testing moratorium, respectively.

13 The same demand has been stated in the context of MST (Altmann 2001: Chs 7, 8).

14 Some of this has been discussed by Altmann (2003).

15 E.g. the prohibition on land-based cruise missiles above 500 km range of the INF Treaty of 1987 applies only to the USA and Russia – all other countries are free in this respect.

16 Note that members of the German delegation at the CFE negotiations explicitly stated: 'The definition of combat aircraft covers all types and variants of combat aircraft that exist or are under development at present, with respect to both, manned as well as eventual future unmanned types' (Hartmann *et al.* 1992: 66, translation: author). I thank Hans-Joachim Schmidt of PRIF for indicating this conscious decision of the negotiators.
17 Heavy armament combat vehicle: ≥6.0 metric tonnes, gun ≥75 mm; armoured infantry fighting vehicle: primarily to transport a combat infantry squad, cannon ≥20 mm, sometimes antitank missile launcher; armoured personnel carrier: transport combat infantry squad, weapon ≤20 mm as a rule (CFE 1990: Art. II).
18 See http://www.robocup.org; http://avdil.gtri.gatech.edu/AUVS/IARCLaunchPoint.html; http://www.engr.arizona.edu/MAVcompetition.
19 This wording includes animals with control electrodes etc., but excludes genetically modified ones. Modifications for the production of toxins are banned by the Biological and Toxin Weapons Convention. At present, achieving control of situational behaviour only by means of genetics seems excluded. Should purely genetic conversion of an animal to a controllable robot become possible, an additional interpretation of such animal as 'artificial' might be needed.
20 The same demand has been stated in the context of MST (Altmann 2001: Chs 7, 8).
21 If mini-/micro-robots were not banned, a clarifying interpretation should be concluded that they count as mines under the Anti-Personnel Mine Convention if they are equipped with explosive and a person sensor.
22 In general, space weapons could be deployed on the ground, in the air or in space, for attacking targets on the ground, in the air or in space. Small satellites would be most dangerous as anti-satellite weapons.
23 Soviet and US anti-satellite systems have been tested, but not deployed, see e.g. Krepon 2001.
24 See Altmann and Scheffran 2003. The same demand has been stated in the context of MST (Altmann 2001: Chs 7, 8).
25 An example of mostly unfounded suspicion is provided by the warning against commercial small satellites and alleged Chinese plans as threats to US satellites in the context of the US Commission on National Security Space (Rumsfeld-II commission) (Wilson 2001: 18, 29 ff.).
26 Strict regulation with intense monitoring and inspection rights would be needed, maybe beyond the present interpretation of privacy. The Foresight Guidelines on MNT (e.g.: legal liability, criminal prosecution where appropriate, no self-replication in a natural environment; evolution is discouraged, distribution of development capability only to responsible actors, encryption of replication information, safety and security measures) give useful general hints (Foresight 2000). They assume that MNT should not be stopped, but recommend substantial research and detailed balancing of benefits and risks.
27 Here the Foresight Guidelines (Foresight 2000) are inconsistent by assuming compliance within states while rejecting international limitation because full 100 per cent verification cannot be achieved, see Section 1.5.6.

7 CONCLUSIONS AND RECOMMENDATIONS

1 Note that there was only one contribution devoted to risks from military uses (Altmann and Gubrud 2002).
2 An example is the termination of work on infrasound acoustic weapons by the US Joint Non-Lethal Weapons Directorate in 1999 due to the lack of a reliable effect (Altmann 2001a).

3 An example is the actual deployment by the USA of ballistic-missile defence interceptors in Alaska (Gronlund *et al.* 2004).

4 At least for 'non-lethal', e.g. incapacitating, agents; for the problems with these see CBWCB 2003; Wheelis 2003; Sunshine 2004.

5 The situation with the Anti-personnel Mine Convention of 1997 is somewhat similar.

BIBLIOGRAPHY

AFOSR (2002) 'Research Interests of the Air Force Office of Scientific Research and Broad Agency Announcement 2003-1', October 2002. Online, available at: www.afosr.af. mil/pages/BAA2003A.htm (accessed 23 April 2003).

—— (2003) Air Force Office of Scientific Research, 'AFRL Directorates'. Online, available at: www.afosr.af.mil/afrdir.htm (accessed 15 April 2003).

Altmann J. (1986) *Laserwaffen – Gefahren für die strategische Stabilität und Möglichkeiten der vorbeugenden Rüstungsbegrenzung*, Schriftenreihe des Arbeitskreises Marburger Wissenschaftler für Friedens- und Abrüstungsforschung Nr. 2, Marburg: Fachbereich Physik der Philipps-Universität (shortened: HSFK-Report 3/1986, Frankfurt/M.: Hessische Stiftung Friedens- und Konfliktforschung).

—— (1994) 'Verifying Limits on Research and Development – Case Studies: Beam Weapons, Electromagnetic Guns', in J. Altmann, T. Stock and J.-P. Stroot (eds) *Verification After the Cold War – Broadening the Process*, Amsterdam: VU Press.

—— (2000) 'Zusammenhang zwischen zivilen und militärischen Hochtechnologien am Beispiel der Luftfahrt in Deutschland', in J. Altmann (ed.) *Dual-use in der Hochtechnologie – Erfahrungen, Strategien und Perspektiven in Telekommunikation und Luftfahrt*, Baden-Baden: Nomos.

—— (2001) *Military Applications of Microsystem Technologies – Dangers and Preventive Arms Control*, Münster: agenda.

—— (2001a) 'Non-lethal Weapons Technologies – the Case for Independent Scientific Analysis', *Medicine, Conflict and Survival* 17 (3): 234–247.

—— (2003) 'Roboter für den Krieg?', *Wissenschaft und Frieden* 21 (3): 18–21.

—— (2004) 'Military Uses of Nanotechnology: Perspectives and Concerns', *Security Dialogue* 35 (1): 61–79.

Altmann J. and Gubrud M. (2002) 'Risks from Military Uses of Nanotechnology – the Need for Technology Assessment and Preventive Control', in M. Roco and R. Tomellini (eds) *Nanotechnology – Revolutionary Opportunities and Societal Implications*, Luxembourg European Communities. Online, available at: www.ep3.ruhr-uni-bochum.de/bvp/RiskMilNT_Leece.pdf (accessed 30 Aug. 2002).

Altmann J. and Scheffran J. (2003) 'New Rules in Outer Space: Options and Scenarios', *Security Dialogue* 34 (1): 109–116.

Andrievski R.A. (2003) 'Modern Nanoparticle Research in Russia', *Journal of Nanoparticle Research* 5 (5–6): 415–418.

Anton P.S., Silberglitt R. and Schneider J. (2001) *The Global Technology Revolution – Bio/Nano/Materials Trends and Their Synergies with Information Technologies by 2015*, Santa Monica CA: RAND.

APMC (1997) 'Convention on the Prohibition of the Use, Stockpiling, Production and Transfer of Anti-Personnel Mines and on their Destruction, 18 Sept. 1997'. Online, available at: www.mines.gc.ca/english/documents/treaty.html (accessed 19 Feb. 2001).

ARL (2002) 'Nanomaterials Research'. Online, available at: www.arl.army.mil/wmrd/Tech/nanoboth.pdf (accessed 24 Jan. 2003).

Armstrong R.E. and Warner J.B. (2003) 'Biology and the Battlefield', *Defense Horizons*, no. 25. Online, available at: www.ndu.edu/inss/DefHor/DH25/DH25.pdf (accessed 8 Dec. 2003).

Army News (2002) 'Army Teams with MIT to Establish ISN', U.S. Army News Release, March 13, 2002. Online, available at www.dtic.mil/armylink/news/Mar2002/r20020313r-02-011.html (accessed 15 March 2002).

Army RDT&E (2001) U.S. Army, *Supporting Data for the FY 2002 Amended President's Budget Submitted to Congress – July 2001, Descriptive Summaries of the RDT&E Army Appropriations, Budget Activities 1, 2, and 3.* Online, available at: www.asafm.army.mil/budget/fybm/FY02/rforms/vol1/vol1.pdf (accessed 26 Feb. 2003).

Arnall A.H. (2003) *Future Technologies, Today's Choices – Nanotechnology, Artificial Intelligence and Robotics: A Technical, Political and Institutional Map of Emerging Technologies*, London: Greenpeace Environmental Trust. Online, available at www.greenpeace.org.uk/Multimedia/Live/FullReport/5886.pdf (accessed 4 Aug. 2003).

Arnett E. (1999) 'Military Research and Development', pp. 351–370 in *SIPRI-Yearbook 1999 – World Armaments and Disarmament*, Stockholm/Oxford: SIPRI/Oxford University Press.

ARO Displays (2001) 'Nanoscience for the Soldier, sub-group on Detectors, Antennae, Displays', in Proceedings from the Workshop on Nanoscience for the Soldier, sponsored by the Army Research Office, February 8–9, 2001, held at the North Carolina Biotechnology Center, Durham NC. Online, available at: www.aro.army.mil/phys/Nanoscience/sec4nano.htm (accessed 10 Sept. 2002).

ARO Materials (2001) 'Materials Group Worksheets', in Proceedings from the Workshop on Nanoscience for the Soldier, sponsored by the Army Research Office, February 8–9, 2001, held at the North Carolina Biotechnology Center, Durham NC. Online, available at: www.aro.army.mil/phys/Nanoscience/sec4material1.html (accessed 10 Sept. 2002).

ARO Nanoscience (2001) Proceedings from the Workshop on Nanoscience for the Soldier, sponsored by the Army Research Office, February 8–9, 2001, held at the North Carolina Biotechnology Center, Durham NC. Online, available at: www.aro.army.mil/phys/Nanoscience (accessed 10 Sept. 2002).

ARO Power (2001) 'Power and Cooling', in Proceedings from the Workshop on Nanoscience for the Soldier, sponsored by the Army Research Office, February 8–9, 2001, held at the North Carolina Biotechnology Center, Durham NC. Online, available at: www.aro.army.mil/phys/Nanoscience/sec4power.htm (accessed 10 Sept. 2002).

ARO Soldier (2001) 'Soldier Status Monitoring', in Proceedings from the Workshop on Nanoscience for the Soldier, sponsored by the Army Research Office, February 8–9, 2001, held at the North Carolina Biotechnology Center, Durham NC. Online, available at: www.aro.army.mil/phys/Nanoscience/sec4soldier.htm (accessed 10 Sept. 2002).

ARO Solicitation (2001) US Army Research Office, 'Broad Agency Announcement, Institute for Soldier Nanotechnologies', October 2001. Online, available at: www.aro.army.mil/soldiernano/finalsolicit.pdf (accessed 20 Jan. 2003).

Ashley S. (2001) 'Nanobot Construction Crews', *Scientific American* 285 (3): 84–85.

Bachmann G. and Zweck A. (2001) 'Nanotechnologie und ihre Folgen', *Wechsel-Wirkung* 23 (110): 36–42.

Bai C. (2001) 'Progress of Nanoscience and Nanotechnology in China', *Journal of Nanoparticle Research* 3 (4): 251–256.

Bainbridge W.S. (2003) 'Challenge and Response', speech at awards ceremony of the Transvision 2003 conference, July 9, 2003 (conference: 25–29 June, 2003). Online, available at: www.transhumanism.com/articles_print.php?id=P697_0_4_0_C, mysite.verizon.net/william.bainbridge/ (accessed 5 Aug. 2003).

Barletta M., Sands A. and Tucker J.B. (2002) 'Keeping Track of Anthrax: The Case for a Biosecurity Convention', *Bulletin of the Atomic Scientists* 58 (3): 57–62.

BICC (2001) Bonn International Center for Conversion, *conversion survey 2001 – Global Disarmament, Demilitarization and Demobilization*, Baden-Baden: Nomos.

Bhushan B. (ed.) (2004) *Springer Handbook of Nanotechnology*, Berlin etc.: Springer.

Bishop F. (1997) 'Some Novel Space Propulsion Systems', draft paper for a talk at the Fifth Foresight Conference on Molecular Nanotechnology. Online, available at: www.foresight.org/Conferences/MNT05/Papers/Bishop (accessed 16 Oct. 2001).

Blumrich R. (1998) 'Technical Potential, Status and Costs of Ground Sensor Systems', in J. Altmann, H. Fischer and H. van der Graaf (eds) *Sensors for Peace – Applications, Systems and Legal Requirements for Monitoring in Peace Operations*, UN Institute for Disarmament Research, New York/Geneva: UNO.

BMBF (1998) *Nanotechnologie – Innovationsschub aus dem Nanokosmos*, Bonn: Bundesministerium für Bildung, Wissenschaft, Forschung und Technologie.

—— (2002) *Nanotechnologie in Deutschland – Standortbestimmung*, Bonn: Bundesministerium für Bildung und Forschung.

Bostrom N. (2002) 'Existential Risks – Analyzing Human Extinction Scenarios and Related Hazards', *Journal of Evolution and Technology* 9. Online, available at: www.jetpress.org/volume9/risks.htm (accessed 7 Nov. 2003).

Brauch H.G., van der Graaf H., Grin J. and Smit W. (1997) *Militärtechnikfolgenabschätzung und präventive Rüstungskontrolle – Institutionen, Verfahren und Instrumente*, Münster/Hamburg: Lit.

Brendley K. and Steeb R. (1993) *Military Applications of Microelectromechanical Systems*, MR-175-OSD/AF/A, Santa Monica CA: RAND.

Brooks R.A. (2002) *Flesh and Machines: How Robots Will Change Us*, New York: Pantheon.

BRTF (2003) Better Regulation Task Force, *Scientific Research: Innovation with Controls*. Online, available at: www.brtf.gov.uk/taskforce/reports/Scientificresearch.pdf (accessed 24 Jan. 2003).

Brumfiel G. (2003) 'A Little Knowledge...', *Nature* 424 (6946): 246–248.

Brzoska M. (2000) 'Sozialwissenschaftliche Forschung zum Dual-use in der Luftfahrt', in J. Altmann (ed.) *Dual-use in der Hochtechnologie – Erfahrungen, Strategien und Perspektiven in Telekommunikation und Luftfahrt*, Baden-Baden: Nomos.

BTWC (1972) 'Convention on the Prohibition of the Development, Production and Stockpiling of Bacteriological (Biological) and Toxin Weapons and on Their Destruction'. Online, available at: www.opcw.org/html/db/cwc/more/biotox.html (accessed 5 Feb. 2004).

Burgess D.E. (2002) 'UK MoD's Nanotechnology Initiatives', viewgraphs presented at *Defence Nanotechnology 2002*, 31 Oct.–1 Nov., London: Defence Event Management.

Busbee J. (2002) 'NanoStructured Materials: Opportunity and Challenge For Aerospace', viewgraphs presented at *Defence Nanotechnology 2002*, 31 Oct.–1 Nov., London: Defence Event Management.

BWC AHG (2001) 'Rolling Text of a Protocol to the Convention on the Prohibition of the Development, Production and Stockpiling of Bacteriological (Biological) and Toxin Weapons and on their Destruction', Geneva, BWC/Ad Hoc Group/56-1, Appendices,.../56-2 (18 May). Online, available at: www.brad.ac.uk/acad/sbtwc/doc56-1.pdf,.../doc56-2.pdf (accessed 9 Feb. 2004).

CBWCB (2003) '"Non-lethal" Weapons, the CWC and the BWC', Editorial, *The CBW Conventions Bulletin*, no. 61. Online, available at: www.sussex.ac.uk/spru/hsp/cbwcb61.pdf (accessed 5 April 2004).

CEC (2000) 'Communication from the Commission on the precautionary principle', COM(2000) 1, Brussels, 2 Feb. 2000, Commission of the European Communities. Online, available at, e.g.: europa.eu.int/comm/dgs/health_consumer/library/pub07_en.pdf (accessed 5 April 2004).

CFE (1990) 'Treaty on Conventional Armed Forces in Europe'. Online, available at: www.state.gov/www/global/arms/treaties/cfe.html, .../cfeadapt.html, .../cfe-1.html, .../cfecomm.html (accessed 24 Oct. 2003).

Chen Y., Jung G.-Y., Ohlberg D.A.A., Li X., Stewart D.R., Jeppesen J.O., Nielsen K.A., Stoddart J.F. and Williams R.S. (2003) 'Nanoscale Molecular-switch Crossbar Circuits', *Nanotechnology* 14 (4): 462–468.

Christaller T., Decker M., Gilsbach J.-M., Hirzinger G., Lauterbach K., Schweighofer E., Schweitzer G. and Sturma D. (2001) *Robotik: Perspektiven für menschliches Handeln in der zukünftigen Gesellschaft*, Berlin etc.: Springer.

Christodolou L. (2000) 'Structural Amorphous Metals (SAM)', Pre-Proposal Workshop June 6, 2000. Online, available at: www.darpa.mil/dso/thrust/matdev/sam/presentations/christodolou.pdf (accessed 4 March 2003).

CINT (2003) 'Center for Integrated Nanotechnologies'. Online, available at: cint.lanl.gov (accessed 27 Jan. 2003).

CMS (2002) *Facts & Figures 2002*, Chemistry and Materials Science Directorate, Lawrence Livermore National Laboratory, UCRL-AR-129465-02. Online, available at: www-cms.llnl.gov/facts_figures_02.pdf (accessed 22 April 2003).

—— (2003) *Facts & Figures 2003*, Chemistry and Materials Science Directorate,

Lawrence Livermore National Laboratory, UCRL-AR-129465-03. Online, available at: www-cms.llnl.gov/facts_figures_03.pdf (accessed 22 April 2003).

Cochran T.B., Arkin W.M. and Hoenig M.M. (1984) *Nuclear Weapons Databook, Volume I – U.S. Nuclear Forces and Capabilities*, Cambridge MA: Ballinger.

Colton R. (2003) 'Nanoscience Research in the NRL Chemistry Division'. Online, available at: nanoscience.nrl.navy.mil/files/NRL_Nanosci_Chem_Open_House.zip/. . .pdf (accessed 15 April 2003).

Colvin V.L. (2003) Testimony, U.S. House of Representatives, Committee on Science, April 9, 2003. Online, available at: www.house.gov/science/hearings/full03/apr09/colvin.htm (accessed 5 May 2003).

Commission (1999) The United States Commission on National Security/21st Century, *New World Coming: American Security in the 21st Century – Major Themes and Implications*, Washington DC: The Commission (15 Sept. 1999) [cited after Metz (2000)].

Companó R. and Hullmann A. (2002) 'Forecasting the Development of Nanotechnology with the help of Science and Technology Indicators', *Nanotechnology* 13 (3): 243–247.

Congress (2003) '21st Century Nanotechnology and Development Act', 108th Congress, 1st Session, S.189. Online, available at, e.g.: www.smalltimes.com/smallstage/images/nanobills189.pdf (accessed 15 Dec. 2003).

COSTIND (2002) 'State Commission of Science, Technology, and Industry for National Defense (COSTIND)'. Online, available at: www.nti:org/db/china/costind.htm (accessed 16 Sept. 2002)

Crandall B.C. (ed.) (1996) *Nanotechnology – Molecular Speculation on Global Abundance*, 3rd printing, Cambridge MA/London: MIT Press.

—— (1996a) 'Molecular Engineering', in B.C. Crandall (ed.), *Nanotechnology – Molecular Speculation on Global Abundance*, 3rd printing, Cambridge MA/London: MIT Press.

Crandall B.C. and Lewis J. (1992) *Nanotechnology – Research and Perspectives, Papers from the First Foresight Conference on Nanotechnology*, 3rd printing, Cambridge MA/London: MIT Press.

CRN (2003) Center for Responsible Nanotechnology, 'CRN Research: Current Results – Dangers of Molecular Nanotechnology'. Online, available at: www.crnano.org/dangers.htm; 'CRN Research: Current Results – No Simple Solutions', www.crnano.org/solutions.htm (accessed 25 April 2003).

CWC (1993) 'Convention On the Prohibition of the Development, Production, Stockpiling and Use of Chemical Weapons and on their Destruction'. Online, available at: www.opcw.org/cwc/cwc-eng.htm (accessed 16 Aug. 2001).

DARPA Activity (2003) 'Activity Detection Technology/Tissue-Based Biosensors'. Online, available at: www.darpa.mil/dso/thrust/biosci/adt-tbbs.htm (accessed 5 May 2003).

DARPA AugCog (2003) 'Augmented Cognition'. Online, available at: www.darpa.mil/ipto/research/ac/index.html, . . ./objectives.html (5 May 2003).

DARPA Biocomp (2003) 'Bio-Computation'. Online, available at: www.darpa.mil/ipto/research/biocomp/index.html, . . ./vision.html, . . ./challenges.html (accessed 5 May 2003).

DARPA BioFlips (2003) 'BioFlips'. Online, available at: www.darpa.mil/dso/thrust/biosci/bioflips.htm (accessed 5 May 2003).

DARPA BioMagnetICs (2003) 'Bio-Magnetic Interfacing Concepts (BioMagnetICs)'. Online, available at: www.darpa.mil/dso/thrust/biosci/biomagn.htm (accessed 5 May 2003).

DARPA Biomolecular (2003) 'Biomolecular Motors'. Online, available at: www.darpa.mil/dso/thrust/biosci/biomomo.htm (accessed 5 May 2003).

DARPA BIOS (2003) 'Biological Input/Output Systems'. Online, available at: www.darpa.mil/dso/thrust/biosci/bios.htm (accessed 29 April 2003).

DARPA Biosensor (2003) 'Biosensor Technologies'. Online, available at: www.darpa.mil/dso/thrust/biosci/biostech.htm (accessed 5 May 2003).

DARPA BMI (2003) 'Brain Machine Interfaces'. Online, available at: www.darpa.mil/dso/thrust/sp/bmi.htm (accessed 16 Jan. 2003).

DARPA BOSS (2003) 'Bio-Optic Synthetic Lens (BOSS)'. Online, available at: www.darpa.mil/dso/thrust/biosci/boss.htm (accessed 5 May 2003).

DARPA Budget (2002) *Fiscal Year 2003 Budget Estimates, Feb. 2002, Research, Development, Test and Evaluation, Defense-Wide, Vol. 1 – Defense Advanced Research Projects Agency*. Online, available at: www.darpa.mil/body/pdf/FY03BudEst.pdf (accessed 4 Nov. 2002).

—— (2003) *Fiscal Year (FY) 2004/FY 2005 Biennial Budget Estimates, Feb. 2003, Research, Development, Test and Evaluation, Defense-Wide, Vol. 1 – Defense Advanced Research Projects Agency*. Online, available at: www.darpa.mil/body/pdf/FY04_FY05BiennialBudgetEstimatesFeb03.pdf (accessed 16 Feb. 2004).

DARPA CAP (2003) 'Continuous Assisted Performance (CAP)'. Online, available at: www.darpa.mil/dso/thrust/biosci/cap.htm (accessed 29 April 2003).

DARPA CBS (2003) 'Controlled Biological and Biomimetic Systems'. Online, available at: www.darpa.mil/dso/thrust/biosci/cbs.htm, .../biosci/cbs/overview.html, .../biosci/cbs/cprogram.html (accessed 13 Feb. 2004).

DARPA Countermeasures (2003) 'Unconventional Pathogen Countermeasures'. Online, available at: www.darpa.mil/dso/thrust/biosci/upathcm.htm (accessed 5 May 2003).

DARPA CSAC (2003) 'CSAC – Chip-Scale Atomic Clock'. Online, available at: www.darpa.mil/mto/csac/overview/index.html (accessed 5 May 2003).

DARPA Diagnostics (2003) 'Advanced Diagnostics'. Online, available at: www.darpa.mil/dso/thrust/biosci/advdiagn.htm (accessed 5 May 2003).

DARPA Energy (2002) 'Energy Harvesting – Projects & Accomplishments, Implantable Biofuel Cell Electrodes'. Online, available at: www.darpa.mil/dso/trans/energy/pa_uta.html (accessed 28 April 2003).

—— (2003) 'Advanced Energy Technologies'. Online, available at: www.darpa.mil/dso/thrust/matdev/advancet.htm (accessed 5 May 2003).

DARPA ETC (2003) 'Engineered Tissue Constructs'. Online, available at: www.darpa.mil/dso/thrust/biosci/etc.htm (accessed 29 April 2003).

DARPA Fact (2003) *Fact File – A Compendium of DARPA Programs*, August 2003. Online, available at: www.darpa.mil/body/pdf/FINAL2003FactFilerev1.pdf (accessed 26 Feb. 2004).

DARPA Genomics (2003) 'Pathogen Genomic Sequencing'. Online, available at: www.darpa.mil/dso/thrust/biosci/genomics.htm (accessed 5 May 2003).

DARPA MARS (2003) 'Mobile Autonomous Robot Software (MARS)'. Online, available at: www.darpa.mil/ipto/research/mars/index.html, .../vision.html (accessed 5 May 2003).

DARPA MEMS (2003) 'DARPA MEMS – Project Summaries'. Online, available at: www.darpa.mil/mto/mems/summaries/Projects/index.html (accessed 5 May 2003), . . ./summaries/projects/university_44.html (accessed 14 Jan. 2003).

DARPA Metabolic (2003) 'BAA03-02, Addendum 2, Special Focus Area: Metabolic Dominance'. Online, available at: www.darpa.mil/baa/baa03-02mod2.htm (accessed 5 May 2003).

DARPA MetaEng (2003) 'Metabolic Engineering for Cellular Stasis'. Online, available at: www.darpa.mil/dso/thrust/biosci/metaeng.htm (accessed 29 April 2003).

DARPA MetaMaterials (2003) 'MetaMaterials'. Online, available at: www.darpa.mil/dso/thrust/matdev/metamat.htm (accessed 5 May 2003).

DARPA MOLDICE (2003) 'Engineered Bio-Molecular Nano-Devices/Systems (MOLDICE)'. Online, available at: www.darpa.mil/dso/future/moldice (accessed 5 May 2003).

DARPA Mole (2003) 'Molecular Electronics (Moletronics)'. Online, available at: www.darpa.mil/MTO/mole (accessed 5 May 2003).

DARPA MOSAIC (2003) 'Molecular Observation, Spectroscopy and Imaging using Cantilevers (MOSAIC)'. Online, available at: www.darpa.mil/dso/thrust/biosci/mosaic.htm (accessed 5 May 2003).

DARPA MPG (2003) 'MPG – Micro Power Generation'. Online, available at: www.darpa.mil/mto/mpg (accessed 5 May 2003).

DARPA QuIST (2003) 'Quantum Information Science and Technology (QuIST)'. Online, available at: www.darpa.mil/dso/thrust/math/quist.htm (accessed 14 April 2003).

DARPA SAM (2003) 'Structural Amorphous Metals (SAM)'. Online, available at: www.darpa.mil/dso/thrust/matdev/sam.htm, 'Presentations', www.darpa.mil/dso/thrust/matdev/sam/html/presentations.html (accessed 26 Feb. 2003).

DARPA SDR (2003) 'Software for Distributed Robotics (SDR)'. Online, available at: www.darpa.mil/ipto/research/sdr/index.html, . . ./vision.html (accessed 5 May 2003).

DARPA SIMBIOSYS (2003) 'Simulation of Bio-Molecular Microsystems (SIMBIOSYS)'. Online, available at: www.darpa.mil/dso/thrust/biosci/simbios.htm (accessed 5 May 2003).

DARPA Smart (2003) 'Smart Materials and Structures'. Online, available at: www.darpa.mil/dso/thrust/matdev/md_smas.htm (accessed 5 May 2003).

DARPA SPINS (2003) 'SPins IN Semiconductors (SPINS)'. Online, available at: www.darpa.mil/dso/thrust/matdev/spins.htm (accessed 5 May 2003).

DARPA Spintronics (2003) 'Spin Transport Electronics (Spintronics)'. Online, available at: www.darpa.mil/dso/thrust/matdev/spintron.htm (accessed 5 May 2003).

DARPA SPSS (2003) 'Small Scale Propulsion Systems'. Online, available at: www.darpa.mil/tto/Programs/ssps.html (accessed 25 April 2003).

DARPA SUNY (2003) 'Development of Biomimetic Robots and Sensors Using Hybrid Brain-Machine Technology', SUNY Health Science Center at Brooklyn. Online, available at: www.darpa.mil/dso/thrust/biosci/cbs/suny_ab.html (accessed 13 Feb. 2004).

DARPA TTO (2003) 'Unmanned Systems – Tactical Multipliers – Space'. Online, available at: www.darpa.mil/tto/programs.html (accessed 6 May 2003).

Devine M. (2002) 'Manufacturing, Research, Development, and Education (RDE) Center for Nanotechnologies', viewgraphs presented at *Defence Nanotechnology 2002*, 31 Oct.–1 Nov., London: Defence Event Management.

De Wild M., Berner S., Suzuki H., Ramoino L, Baratoff A. and Jung T.A. (2003) 'Molecular Assembly and Self-Assembly – Molecular Nanoscience for Future Technologies', *Annals of the New York Academy of Sciences* 1006 (1): 291–305.

DoD (2001) Department of Defense, *Space Technology Guide, FY 2000–01*, Office of the Secretary of Defense. Online, available at: www.space.gov/technology/myer.pdf (accessed 29 Jan. 2002).

DoD fellowships (2001) 'National Nanotechnology Initiative, Activities: DoD Fellowships in Nanoscience for FY01'. Online, available at: www.nano.gov/durintfellow01.htm (accessed 20 Jan. 2003).

DoE (2003) 'Nanoscale Science, Engineering, and Technology Research (NSET)'. Online, available at: www.sc.doe.gov/bes/NNI.htm (accessed 22 April 2003).

Dowding R.J. (2003) 'Materials for Kinetic Energy Penetrators', Structural Amorphous Metals (SAM) Pre-Proposal Conference, June 6, 2000, Arlington, VA. Online, available at: www.darpa.mil/dso/thrust/matdev/sam/html/presentations/bdowding.pdf (accessed 26 Feb. 2003).

Drexler K.E. (1981) 'Molecular Engineering: An Approach to the Development of General Capabilities for Molecular Manipulation', *Proceedings of the National Academy of Sciences* 78 (9): 5275–5278.

—— (1986/1990) *Engines of Creation – The Coming Era of Nanotechnology*, New York: Anchor/Doubleday.

—— (1992) *Nanosystems – Molecular Machinery, Manufacturing, and Computation*, New York etc.: Wiley.

Drexler E. and Peterson C. (with Gayle P.) (1991) *Unbounding the Future: the Nanotechnology Revolution*, New York: Morrow. Online, available at: www.foresight.org/UTF/Unbound_LBW/index.html (accessed 17 July 2001).

Drexler K.E. and Smalley R.E. (2003) 'Nanotechnology – Drexler and Smalley make the case for and against "molecular assemblers"', *Chemical & Engineering News* 82 (48): 37–42.

Drexler K.E., Forrest D., Freitas R.A., Hall J.S., Jacobstein N., McKendree T., Merkle R. and Peterson S. (2001) 'On Physics, Fundamentals, and Nanorobots: A Rebuttal to Smalley's Assertion…', *Foresight Update* no. 46: 6–7 (30 Sept.). Online, available at: www.imm.org/SciAmDebate2/smalley.html (accessed 4 Nov. 2003).

—— (2001a) 'Many Future Nanomachines: A Rebuttal to Whiteside's Assertion…', *Foresight Update* no. 46: 8–17 (30 Sept.). Online, available at: www.imm.org/SciAmDebate2/whitesides.html (accessed 4 Nov. 2003).

DURINT (2001) 'The DoD Fiscal Year 2001, Defense University Initiative on Nanotechnology'. Online, available at: www.onr.navy.mil/sci_tech/special/durint/durint01baa.htm (accessed 17 July 2001).

—— (2001a) 'Nanotechnology Research Awards Announced'. Online, available at: www.defenselink.mil/news/Feb2001/b02232001_bt079-01.html (accessed 17 Sept. 2002).

—— (2001b) 'FY 2001 DURINT Equipment'. Online, available at: www.defenselink.mil/news/Feb2001/d20010223equip.pdf (accessed 17 July 2001).

—— (2001c) 'FY 2001 DURINT'. Online, available at: www.defenselink.mil/news/Feb2001/d20010223durint.pdf (accessed 17 July 2001).

EC (2004) Commission of the European Communities, 'On the Implementation of the Preparatory Action on the Enhancement of the European Industrial Potential in the Field of Security Research, Towards a Programme to Advance European Security through Research and Technology', COM(2004) 72 final, Brussels, 3 Feb. 2004. Online, available at: europa.eu.int/eur-lex/en/com/cnc/2004/comm2004_0072en01.pdf (accessed 5 April 2004).

Edwards B.C. (2000) *The Space Elevator*, Manuscript for NASA Institute for Advanced Concepts. Online, available at: www.niac.usra.edu/files/studies/final_report/pdf/472Edwards.pdf (accessed 13 Dec. 2002).

—— (2000a) 'Design and Deployment of a Space Elevator', *Acta Astronautica* 47 (19): 735–744.

EML (2003) *Selected Papers from the 11th Symposium on Electromagnetic Launch Technology*, IEEE Transactions on Magnetics 39 (1).

ENMOD (1977) 'Convention on the Prohibition of Military or Any Other Hostile Use of Environmental Modification Techniques'. Online, available at: www.opcw.org/html/db/cwc/more/enmod.html (accessed 5 Feb. 2004).

ETC (2003) *From Genomes to Atoms – The Big Down – Atomtech: Technologies Converging at the Nanoscale*, Winnipeg: ETC Group. Online, available at: www.etcgroup.org/documents/TheBigDown.pdf (accessed 3 Feb. 2003).

—— (2003a) *No Small Matter II: The Case for a Global Moratorium – Size Matters!*, ETC Group Occas. Paper Series 7 (1). Online, available at: www.etcgroup.org/documents/Occ_Paper_Nanosafety.pdf (accessed 25 April 2003).

EU FP6 NMP (2003) 'Nanotechnology and Nanosciences, Knowledge-Based Multifunctional Materials, New Production Processes and Devices'. Online, available at: www.cordis.lu/nmp/home.html (accessed 4 Nov. 2003).

Evison D., Hinsley D. and Rice P. (2002) 'Chemical Weapons', *British Medical Journal* 324 (7333): 332–335.

Feakes D. and Littlewood J. (2002) 'Hope and Ambition turn to Dismay and Neglect: The Biological and Toxin Weapons Convention in 2001', *Medicine, Conflict and Survival* 18 (2): 161–174.

Feynman R. (1959) 'There's Plenty of Room at the Bottom: An Invitation to Enter a New Field of Physics', speech, 29 December 1959, American Physical Society Annual Meeting, California Institute of Technology, originally published: Engineering and Science (CalTech), Feb. 1960: 22–36. Online, available at, e.g.: www.its.caltech.edu/~feynman, www.zyvex.com/nanotech/feynman.html (accessed 25 Aug. 2003).

Foresight (1995) *Fourth Foresight Conference on Molecular Nanotechnology*, 1995. Online, available at: www.zyvex.com/nanotech/nano4.html (accessed 4 Nov. 2003).

—— (1995a) 'Space Development Advocates Support Nanotechnology', *Foresight Update* no. 20: (5). Online, available at: www.foresight.org/Updates/Update20/Update20.5.html (accessed 16 Oct. 2001).

—— (2000) 'Foresight Guidelines on Molecular Nanotechnology', Revised Draft Version 3.7: June 4, 2000. Online, available at: www.foresight.org/guidelines/current.html (accessed 14 April 2003).

—— (2002) *Foresight Update* no. 48: 2. Online, available at: www.foresight.org/Updates/Update48/Update48.2.html (accessed 4 Nov. 2003).

—— (2003) 'Nanotechnology Prizes and Awards Sponsored by Foresight'. Online, available at: www.foresight.org/FI/fi_spons.html (accessed 25 Aug. 2003).

—— (2003a) 'Foresight Nanotechnology FAQ – About Foresight Institute'. Online, available at: www.foresight.org/NanoRev/FIFAQ3.html (accessed 3 Nov. 2003).

Frackowiak E. and Béguin F. (2002) 'Electrochemical Storage of Energy in Carbon Nanotubes and Nanostructured Carbons', *Carbon* 40 (10): 1775–1787.

Franks A. (1987) 'Nanotechnology', *Journal of Physics E: Scientific Instruments* 20 (12): 1442–1451.

Freitas R.A. (1996/99) 'Respirocytes – A Mechanical Artificial Red Cell: Exploratory Design in Medical Nanotechnology'. Online, available at: www.foresight.org/Nanomedicine/Respirocytes.html (accessed 18 Aug. 2003).

—— (1999) *Nanomedicine, Vol. I: Basic Capabilities*, Georgetown TX: Landes Bioscience.

—— (2000) 'Some Limits to Global Ecophagy by Biovorous Nanoreplicators, with Public Policy Recommendations'. Online, available at: www.foresight.org/NanoRev/Ecophagy.html (accessed 20 July 2001).

Geiger G. (2003) *Rüstungspotentiale neuer Mikrotechnologien – Konsequenzen für internationale Sicherheit und Rüstungskontrolle*, SWP-Studie S 24, Berlin: Stiftung Wissenschaft und Politik.

Glaser A. (2002) 'The Conversion of Research Reactors to Low-Enriched Fuel and the Case of the FRM-II', *Science and Global Security* 10 (1): 61–79.

Glasstone S. and Dolan P.J. (1977) *The Effects of Nuclear Weapons*, Washington DC: US Department of Defense and Energy Research and Development Administration/US Government Printing Office.

Goddard III W.A., Brenner D.W., Lyshevski S.E. and Iafrate G.J. (eds) (2002) *Handbook of Nanoscience, Engineering, and Technology*, Boca Raton FL: CRC Press.

Goronkin H., von Allmen P., Tsui R.K. and Zhu T.X. (1999) 'Functional Nanoscale Devices', in R.W. Siegel, E. Hu and M.C. Roco (eds) *Nanostructure Science and Technology – A Worldwide Study*, Baltimore MD: WTEC, Loyola College. Also: Boston etc.: Kluwer. Online, available at: www.wtec.org/loyola/pdf/nano.pdf (accessed 25 April 2003).

Gracias D.H., Tien J., Breen T.L., Hsu C. and Whitesides G.M. (2000) 'Forming Electrical Networks in Three Dimensions by Self-Assembly', *Science* 298 (5482): 1170–1172.

Gronlund L., Wright D.C., Lewis G.N. and Coyle III P.E. (2004) *Technical Realities – An Analysis of the 2004 Deployment of a U.S. National Missile Defense System*, Cambridge MA: Union of Concerned Scientists.

Gross M. (1999) *Travels to the Nanoworld – Miniature Machinery in Nature and Technology*, Cambridge MA: Perseus.

Gsponer A. (2002) 'From the Lab to the Battlefield? Nanotechnology and Fourth-Generation Nuclear Weapons', *Disarmament Diplomacy*, no. 67. Online, available at: www.acronym.org.uk/dd/dd67/67op1.htm (accessed 11 Nov. 2002).

Gsponer A. and Hurni J.-P. (2000) *The Physical Principles of Thermonuclear Explosives, Inertial Confinement Fusion, and the Quest for Fourth Generation Nuclear Weapons*, 7th edn, INESAP Technical Report no. 1, Darmstadt: IANUS, Darmstadt University of Technology.

Gubrud M. (1989) 'Nanotechnology in Warfare', *Foresight Update* no. 5 (1 March 1989). Online, available at: www.foresight.org/Updates/Update05/Update05.2.html (accessed 23 Jan. 2001).

—— (1997) 'Nanotechnology and International Security', paper presented at 5th Foresight Conference on Molecular Nanotechnology. Online, available at: www.foresight.org/Conferences/MNT05/Papers/Gubrud (accessed 31 Oct. 2000).

Haberzettl C.A. (2002) 'Nanomedicine: destination or journey?', *Nanotechnology* 13 (4): R9–R13.

Hall J.S. (1999) 'Architectural considerations for self-replicating manufacturing systems', *Nanotechnology* 10 (3): 323–330.

Hall S.S. (2003) 'The Quest for a Smart Pill', *Scientific American* 289 (3): 54–65.

Hanson R. (1998) 'A Critical Discussion of Vinge's Singularity Concept'. Online, available at: www.extropy.org/ideas/journal/previous/1998/10-01.html (accessed 23 June 2003).

Harper T. (2002) 'The nanotechnology arms race: why nobody wants to be left behind', 14 Nov. 2002. Online, available at: www.nanotechweb.org/articles/column/1/11/1/1 (accessed 25 Aug. 2003).

—— (2003) 'What is Nanotechnology?', *Nanotechnology* 14 (1): [no p.].

Hartmann R., Heydrich W. and Meyer-Landrut N. (1992) *Kommentar zum Vertrag über Konventionelle Streitkräfte in Europa*, SWP-S 375, Ebenhausen: Stiftung Wissenschaft und Politik.

Heller A. (1998) 'Keeping Laser Development on Target for the National Ignition Facility', *Science & Technology Review* (March): 4–13.

—— (1999) 'Target Chamber's Dedication Marks a Giant Milestone', *Science & Technology Review* (September): 16–19.

Henley L.D. (1999) 'The RMA After Next', *Parameters* 29: 46–57.

Hoag H. (2003) 'Remote Control', *Nature* 423 (6942): 796–798.

Holdridge G.M. (ed.) (1999) *Russian Research and Development Activities on Nanoparticles and Nanostructured Materials*, Baltimore MD: International Technology Research Institute. Online, available at: wtec.org/loyola/nano/Russia/nanorussia.pdf (accessed 25 April 2003).

House (2003) Committee on Science, U.S. House of Representatives, Hearing Charter, 'The Societal Implications of Nanotechnology', April 9, 2003, Appendix I. Online, available at: www.house.gov/science/hearings/full03/apr09/charter.htm (accessed 5 May 2003).

Houser E.J. and McGill R.A. (2002) 'Chemical Sensor Applications of Nanomaterials', viewgraphs presented at *Defence Nanotechnology 2002*, 31 Oct.–1 Nov., London: Defence Event Management.

Howard C.V. (2003) 'Nano-particles and Toxicity', in *No Small Matter II: The Case for a Global Moratorium – Size Matters!*, ETC Group Occas. Paper Series 7 (1). Online, available at: www.etcgroup.org/documents/Occ_Paper_Nanosafety.pdf (accessed 25 April 2003).

Howard S. (2002) 'Nanotechnology and Mass Destruction: The Need for an Inner Space Treaty', *Disarmament Diplomacy*, no. 65 (July–August). Online, available at: www.acronym.org.uk/dd/dd65/65op1.htm (accessed 26 Aug. 2002).

Hoyle C. and Koch A. (2002) 'Yemen Drone Strike: Just the Start?', *Jane's Defence Weekly* 38 (20): 3.

HRW (2003) 'Human Rights Watch, The United States and the International

Criminal Court'. Online, available at: hrw.org/campaigns/icc/us.htm (accessed 5 April 2004).

IBM (2002) 'IBM's "Millipede" Project Demonstrates Trillion-Bit Data Storage Density'. Online, available at: www.research.ibm.com/resources/news/20020611_millipede.shtml (accessed 26 Sept. 2003).

ICRC (1977) 'Protocols Additional to the Geneva Conventions of 12 August 1949', Geneva, International Committee of the Red Cross. Online, available via: www.icrc.org/ihl.nsf (accessed 5 Feb. 2004).

Infocastinc (2003) 'NBIC Convergence 2003'. Online, available at: www.infocastinc.com/NBIC/nbic.asp (accessed 5 Aug. 2003).

INT (2003) 'Wehrtechnische Implikationen der Nanotechnologie'. Online, available at: www.int.fhg.de, search: Nanotechnologie (accessed 27 Oct. 2003).

IPSE (2003) Internships In Public Science Education at UW-Madison, 'Helping the Public Understand Nanotechnology, Amorphous Metal Activity Guide'. Online, available at: mrsec.wisc.edu/edetc/IPSE/Activities%20HTML/Amorph%20Activity%20Guide.html (accessed 26 Feb. 2003).

ISN (2002) Institute for Soldier Nanotechnologies, 'Mission and Overview'. Online, available at: web.mit.edu/isn/overview.html (accessed 29 Oct. 2002).

ISN 2 (2002) Institute for Soldier Nanotechnologies, 'TEAM 2: Mechanically Active Materials & Devices'. Online, available at: web.mit.edu/isn/research/team2.html (accessed 29 Oct. 2002).

ISN Q&A (2002) Institute for Soldier Nanotechnology (ISN), 'Questions and Answers (provided by the Department of the Army)', MIT News Release, March 13, 2002. Online, available at: web.mit.edu/newsoffice/nr/2002/isnqa.html (accessed 15 March 2002).

ISN Research (2002) Institute for Soldier Nanotechnologies, 'Research'. Online, available at: web.mit.edu/isn/research/index.html (accessed 29 Oct. 2002).

ITRS (2002) *International Technology Roadmap for Semiconductors – 2002 Update*. Online, available at: public.itrs.net/Files/2002Update/2002Update.pdf (accessed 10 Nov. 2003).

—— (2003) *International Technology Roadmap for Semiconductors, 2003 Edition, Executive Summary*, Online, available at: public.itrs.net/Files/2003ITRS/ExecSum2003.pdf (accessed 16 Feb. 2004).

IWGN (1999) *Nanotechnology Research Directions: IWGN Workshop Report*, Washington DC: National Science and Technology Council (Sept. 1999). Online, available at: www.wtec.org/loyola/nano/IWGN.Research.Directions/IWGN-rd.pdf (accessed 10 Sept. 2002).

—— (1999a) *Nanotechnology – Shaping the World Atom by Atom*, Washington DC, National Science and Technology Council (Sept. 1999). Online, available at: www.wtec.org/loyola/nano/IWGN.PublicBrochure/IWGN.Nanotechnology.Brochure.pdf (accessed 16 Nov. 2003).

JDCC (2003) *Strategic Trends*, Shrivenham: Joint Doctrine and Concepts Centre. Online, available at: www.mod.uk/jdcc/trends.htm (accessed 16 Feb. 2004).

Jelinski L. (1999) 'Biologically Related Aspects of Nanoparticles, Nanostructured Materials, and Nanodevices', in R.W. Siegel, E. Hu and M.C. Roco (eds) *Nanostructure Science and Technology – A Worldwide Study*, Baltimore MD: WTEC, Loyola College. Also: Boston etc.: Kluwer. Online, available at: www.wtec.org/loyola/pdf/nano.pdf (accessed 25 April 2003).

Jeremiah D.E. (1995) 'Nanotechnology and Global Security', paper presented at 4th Foresight Conference on Molecular Nanotechnology. Online, available at: www.zyvex.com/nanotech/nano4/jeremiahPaper.html (accessed 14 April 2003).

Jetpress (2003) *Journal of Evolution and Technology*. Online, available at: www.jetpress.org/index.html (accessed 23 June 2003).

Johnson W.L. (1999) 'Bulk Glass-Forming Metallic Alloys: Science and Technology', *MRS Bulletin* 24 (19): 42–56.

Joy B. (2000) 'Why the Future Doesn't Need Us – Our most Powerful 21st-Century Technologies – Robotics, Genetic Engineering, and Nanotech – are Threatening to make Humans an Endangered Species', *Wired* 8.04. Online, available at: www.wired.com/wired/archive/8.04/joy_pr.html (accessed 16 Jan. 2003).

JV (2000) *Joint Vision 2020*, General Henry H. Shelton, Chairman of the Joint Chiefs of Staff, Washington DC: Government Printing Office.

Kaehler T. (1996) 'In-Vivo Nanoscope and the "Two-Week Revolution"', in B.C. Crandall (ed.) *Nanotechnology – Molecular Speculation on Global Abundance*, 3rd printing, Cambridge MA/London: MIT Press.

Kafafi Z.H. (2003) 'Nanoscience Research in Optics'. Online, available at: nanoscience.nrl.navy.mil/files/NANO_RESEARCH_IN_OPTICS.zip, . . .ppt (accessed 15 April 2003).

Killion T.H. (1997) 'Army Basic Research Strategy', *Military Review* 77 (2). Online, available at: www-cgsc.army.mil/milrev/English/MarApr97/Killion.htm (accessed 16 Jan. 2003).

Klaassen C.D., Amdur M.O. and Doull J. (eds) (1986) *Casarett and Doull's Toxicology – The Basic Science of Poisons*, 3rd edn, New York etc.: Macmillan.

Klabunde K. (2000) 'NanoTechnology Solutions to Weapons of Mass Destruction', paper presented at *Non-Lethal Technology and Academic Research Symposium*, NTAR II, 15–17 Nov. 2000, Portsmouth NH. Online, available at: www.unh.edu/ntar/PDF/klabunde.pdf (accessed 28 May 2001).

Koch C. (1999) 'Bulk Behavior of Nanostructured Materials', in R.W. Siegel, E. Hu and M.C. Roco (eds) *Nanostructure Science and Technology – A Worldwide Study*, Baltimore MD: WTEC, Loyola College. Also: Boston etc.: Kluwer. Online, available at: www.wtec.org/loyola/pdf/nano.pdf (accessed 25 April 2003).

Krepon M. (2001) 'Lost in Space: The Misguided Drive Toward Antisatellite Weapons', *Foreign Affairs* 80 (3): 2–8.

Kurzweil R. (1999) *The Age of Spiritual Machines – When Computers Exceed Human Intelligence*, New York etc.: Penguin.

Lanz W., Odermatt W. and Weihrauch G. (2001) 'Kinetic Energy Projectiles: Development History, State of the Art, Trends', 19th International Symposium of Ballistics, 7–11 May 2001, Interlaken, Switzerland. Online, available at: 128.121.234.149/.ttk/symp_19/TB191191.pdf (accessed 10 Feb. 2003).

Lau K.-T. and Hui D. (2002) 'The Revolutionary Creation of Advanced Materials – Carbon Nanotube Composites', *Composites Part B* 33: 263–277.

LDRD (2003) 'Laboratory Directed Research and Development'. Online, available at: www.llnl.gov/llnl/04science/LDRD.html (accessed 22 April 2003).

Lee C.K., Wu M.K. and Yang J.C. (2002) 'A Catalyst to Change Everything: MEMS/NEMS – a Paradigm of Taiwan's Nanotechnology Program', *Journal of Nanoparticle Research* 4 (5): 377–386.

Lee J.-W. (2002) 'Overview of nanotechnology in Korea – 10 years blueprint', *Journal of Nanoparticle Research* 4 (6): 377–386.

Lem S. (1983) *The Upside-Down Evolution*, available in the collection 'One Human Minute', Orlando FL: Harvest/HBJ, 1986.

Leo A. (2002) 'The Soldier of Tomorrow – The U.S. Army Enlists the Massachusetts Institute of Technology to Build the Uniform of the Future'. Online, available at: www.technologyreview.com/articles/leo032002.asp (accessed 8 April 2002).

Libicki M.C. (1994) *The Mesh and the Net – Speculations on Armed Conflict in a Time of Free Silicon*, McNair Paper 29, Washington DC: National Defense University. Online, available at: www.shipwright.com/meshnet.html (accessed 23 Jan. 2001).

Lieber C.M. (2001) 'The Incredible Shrinking Circuit', *Scientific American* 285 (3): 59–64.

Llinas R.R. and Makarov V.A. (2003) 'Brain-Machine Interface via a Neurovascular Approach', in M.C. Roco and W.S. Bainbridge (eds) *Converging Technologies for Improving Human Performance – Nanotechnology, Biotechnology, Information Technology and Cognitive Science*, Boston etc.: Kluwer. Online, available at: www.wtec.org/ConvergingTechnologies/NBIC_report.pdf (accessed 29 Oct. 2003).

Lloyd S. (2000) 'Ultimate physical limits to computation', *Nature* 406 (6799): 1047–1054.

—— (2002) 'Computational Capacity of the Universe', *Physical Review Letters* 88 (23): 237901-1–237901-4.

MacDonald N.C. (1999) 'Nanostructures in Motion: Micro-Instruments for Moving Nanometer-Scale Objects', in G. Timp (ed.) *Nanotechnology*, New York: AIP/Springer.

McKendree T. (1998) 'The logical core architecture', *Nanotechnology* 9 (3): 212–222.

—— (2001) 'Summary of a Dissertation on Molecular Nanotechnology For Space Operations', abstract for presentation at Ninth Foresight Conference on Molecular Nanotechnology. Online, available at: www.foresight.org/conferences/MNT9/Abstracts/McKendree (accessed 16 Oct. 2001).

Magness L., Kecskes L., Chung M., Kapoor D., Biancianello F. and Ridder S. (2001) 'Behavior and Performance of Amorphous and Nanocrystalline Metals in Ballistic Impacts', paper presented at 19th International Symposium of Ballistics, 7–11 May 2001, Interlaken, Switzerland. Online, available at: 128.121.234.149/.ttk/symp_19/TB181183.pdf (accessed 10 Feb. 2003).

Malanowski N. (2001) *Vorstudie für eine Innovations- und Technikanalyse (ITA) Nanotechnologie*, Düsseldorf: VDI-Technologiezentrum (shortened: *Frankfurter Rundschau* 26 May 2001).

Mano N., Mao F. and Heller A. (2002) 'A Miniature Biofuel Cell Operating in A Physiological Buffer', *Journal of the American Chemical Society* 124, 12962–12963.

Merkle R.C. (1989) 'Energy Limits to the Computational Power of the Human Brain', *Foresight Update* no. 6. Online, available at: www.merkle.com/brainLimits.html (accessed 23 Jan. 2001).

—— (1992) 'The Risks of Nanotechnology', in B.C. Crandall and J. Lewis (eds) *Nanotechnology – Research and Perspectives, Papers from the First Foresight*

Conference on Nanotechnology, 3rd printing, Cambridge MA/London: MIT Press.

—— (1997) 'A Proposed "Metabolism" for a Hydrocarbon Assembler', *Nanotechnology* 8 (4): 149–162.

—— (1999) 'Casing an Assembler', *Nanotechnology* 10 (3): 315–322.

—— (2000) 'Molecular Building Blocks and Development Strategies for Molecular Nanotechnology', *Nanotechnology* 11 (2): 89–99.

Metz S. (2000) 'The Next Twist of the RMA', *Parameters* 30 (3): 40–53. Online, available at: carlisle-www.army.mil/usawc/parameters/00autumn/metz.htm (accessed 13 Dec. 2002).

Meyer M. (2001) 'Socio-economic Research on Nanoscale Science and Technology: A European Overview and Illustration', in M.C. Roco and W.S. Bainbridge (eds) *Societal Implications of Nanoscience and Nanotechnology* (NSF Report), Boston etc.: Kluwer. Online, available at: www.wtec.org/loyola/nano/societalimpact/nanosi.pdf (accessed 22 Sept. 2003).

Meyyappan M. and Svrivastava D. (2002) 'Carbon Nanotubes', in W.A. Goddard III, D.W. Brenner, S.E. Lyshevski and G.J. Iafrate G.J. (eds) *Handbook of Nanoscience, Engineering, and Technology*, Boca Raton FL: CRC Press.

Miasnikov E. (2000) 'Precision Guided Weapons and Strategic Balance' (in Russian), Dolgoprudny: Center for Arms Control, Energy and Environmental Studies at MIPT. Online, available at: www.armscontrol.ru/start/publications/vto1100.htm (accessed 18 April 2002).

Minatec (2001) 'Minatec 2001'. Online, available at: www.minatec.com/us/minatec.htm, . . ./minatec.pdf (accessed 18 Jan. 2001).

—— (2003) 'Programme'. Online, available at: www.minatec.com/minatec2003/Minatec2003prog.pdf (accessed 5 April 2004).

MIT News (2002) 'Army Selects MIT for $50 million Institute', MIT News, March 13, 2002, updated March 14, 2002. Online, available at: web.mit.edu/newsoffice/nr/2002/isn.html (accessed 15 March 2002).

MMSG (2000) 'Molecular Manufacturing Shortcut Group – Promote the Development of Nanotechnology as a Means to Facilitate the Settlement of Space'. Online, available at: www.islandone.org/MMSG (accessed 26 Sept. 2003).

Mnyusiwalla A., Daar A.S. and Singer P.A. (2003) ' "Mind the Gap": Science and Ethics in Nanotechnology', *Nanotechnology* 14 (3): R9–R13.

Montemagno C.D. (2001) 'Nanomachines: A Roadmap for Realizing the Vision', *Journal of Nanoparticle Research* 3 (1): 1–3.

Moravec H. (1988) *Mind Children – The Future of Robot and Human Intelligence*, Cambridge MA/London: Harvard University Press.

Moriarty P. (2001) 'Nanostructured Materials', *Reports on Progress in Physics* 64, 297–381.

Morton J.S. (1998) 'The Legal Status of Laser Weapons that Blind', *Journal of Peace Research* 35 (6): 697–705.

Moteff J.D. (1999) *Defense Research: A Primer on the Department of Defense's Research, Development, Test and Evaluation (RDT&E) Program*, CRS Report 97-316. Online, available at: www.NCSEonline.org/nle/crsreports/science/st-63.cmf (accessed 27 Nov. 2002).

MRI (2003) 'Materials Research Institute'. Online, available at: www.cms.llnl.gov/MRI (accessed 22 April 2003).

Mullins W.M. (2002) 'Institute for Soldier Nanotechnologies', viewgraphs presented at *Defence Nanotechnology 2002*, 31 Oct.–1 Nov., London: Defence Event Management.

Murday J.S. (1999) 'Science and Technology of Nanostructures in the Department of Defense', *Journal of Nanoparticle Research* 1 (4): 501–505.

Nalwa H.S. (1999) *Handbook of Nanostructured Materials and Nanotechnology* (5 volumes), San Diego etc.: Academic Press.

NASA (2002) 'Space Settlement Basics'. Online, available at: www.nas.nasa.gov/About/Education/SpaceSettlement/Basics/wwwwh.html (accessed 11 Nov. 2003).

—— (2002a) 'Novel Data Storage System', Center for Nanotechnology, NASA Ames Research Center. Online, available at: www.ipt.arc.nasa.gov/datastorage.html (accessed (16 Sept. 2002).

Natick (1998) 'Up and Atom – Great Things Come in Small Packages'. Online, available at: www.natick.army.mil/warrior/98/nano.htm (accessed 10 Sept. 2002).

Nelson M. and Shipbaugh C. (1995) *The Potential of Nanotechnology for Molecular Manufacturing*, MR-615-RC, Santa Monica CA: RAND.

Netfirms (2001) 'Executive Summary'. Online, available at: www.nanoisrael.netfirms.com/executive_summary.htm (accessed 26 Sept. 2003).

Neuneck G. and Mölling C. (2001) 'Methoden, Kriterien und Konzepte für Präventive Rüstungskontrolle', *Wissenschaft und Frieden*, Dossier Nr. 38.

Neuneck G. and Mutz R. (2000) *Vorbeugende Rüstungskontrolle. Ziele und Aufgaben unter besonderer Berücksichtigung verfahrensmäßiger und institutioneller Umsetzung im Rahmen internationaler Regime*, Baden-Baden: Nomos.

Nicolelis M.A.L. (2001) 'Actions from Thoughts', *Nature* 409 (6818): 403–407.

Nicolelis M.A.L. and Chapin J.K. (2002) 'Controlling Robots with the Mind', *Scientific American* 287 (4): 24–31.

Nixdorff K. (2003) 'Statement on Biosecurity of the INES Working Group', *INESAP Bulletin* no. 22: 39 (December). Online, available at: www.inesap.org/bulletin22/bul22art14.htm (accessed 13 Feb. 2004).

Nixdorff K., Hotz M., Schilling D. and Dando M. (2003) *Biotechnology and the Biological Weapons Convention*, Münster: agenda.

NNI (2000) *National Nanotechnology Initiative: The Initiative and its Implementation Plan*, Washington DC: National Science and Technology Council. Online, available at: nano.gov/nni2.pdf (accessed 17 July 2002).

—— (2002) *National Nanotechnology Initiative: The Initiative and its Implementation Plan*, Washington DC: National Science and Technology Council. Online, available at: nano.gov/nni03_aug02.pdf (accessed 10 Sept. 2002).

—— (2003) *National Nanotechnology Initiative – Research and Development Supporting the Next Industrial Revolution – Supplement to the President's FY 2004 Budget*, Washington DC: National Science and Technology Council. Online, available at: nano.gov/nni04_budget_supplement.pdf (accessed 17 Nov. 2003).

NNI Committee (2002) Committee for the Review of the National Nanotechnology Initiative, *Small Wonders, Endless Frontiers – A Review of the National Nanotechnology Initiative*, Washington DC: National Academy Press.

NRC Committee (2002) Committee on Implications of Emerging Micro- and

Nanotechnologies, Air Force Science and Technology Board, National Research Council, *Implications of Emerging Micro- and Nanotechnologies*, Washington DC: National Academies Press.

NRL Nanoscience (2003) 'Nanoscience Institute'. Online, available at: nrl.navy.mil/content.php?P=MULTIDISCIPLINE (accessed 14 April 2003).

—— (2003a) Institute for Nanoscience, 'Current Topics in the Nanoscience Task Area'. Online, available at: nanoscience.nrl.navy.mil/programs.html (accessed 14 April 2003).

NRL NT Funding (2002) 'Nanoscience and Nanotechnology – DOD Funding Agencies'. Online, available at: nanosra.nrl.navy.mil/funding.html (accessed 28 April 2003).

NRL NT Labs (2002) 'Nanoscience and Nanotechnology – DOD Laboratories'. Online, available at: nanosra.nrl.navy.mil/laboratories.html (accessed 28 April 2003).

NSET (2000) 'Nanotechnology Definition (NSET, February 2000)'. Online, available at: nano.gov/omb_nifty50.htm (accessed 13 Dec. 2002).

NSS (1995) 'NSS Position Paper on Space and Molecular Nanotechnology', National Space Society. Online, available at: www.islandone.org/MMSG/NSS-NanoPosition.html (accessed 16 Oct. 2001).

Olds J. (1958) 'Self-Stimulation of the Brain', *Science* 127 (3294): 315–324.

OTA (1991) U.S. Congress, Office of Technology Assessment, *Miniaturization Technologies*, OTA-TCT-514, Washington DC: U.S. Government Printing Office.

Parker A. (2000) 'New Day Dawns in Supercomputing', *Science & Technology Review* (June), 4–14.

—— (2000a) 'Nanoscale Chemistry Yields Better Explosives', *Science & Technology Review* (October), 19–22.

Paschen H., Coenen C., Fleischer T., Grünwald R., Oertal D. and Revermann C. (2003) *TA-Projekt Nanotechnologie – Endbericht*, TAB Arbeitsbericht Nr. 92, Berlin: Büro für Technikfolgen-Abschätzung beim Deutschen Bundestag, Juli 2003.

Peres S. (2003) 'Nanotechnology holds a Key to Israel's Future', *Jerusalem Post* (6 March 2003). Online, available at: www.smalltimes.com/document_display.cfm?document_id=5613 (accessed 26 Sept. 2003).

Petermann T., Socher M. and Wennrich C. (1997) *Präventive Rüstungskontrolle bei Neuen Technologien – Utopie oder Notwendigkeit?*, Berlin: edition sigma.

Petersen J.L. and Egan D.M. (2002) 'Small Security: Nanotechnology and Future Defense', *Defense Horizons*, no. 8. Online, available at: www.ndu.edu/inss/DefHor/DH8/DH08.pdf (accessed 13 Dec. 2002).

Petro J.B., Plasse T.E. and McNulty J.A. (2003) 'Biotechnology: Impact on Biological Warfare and Biodefense', *Biosecurity and Bioterrorism: Biodefense Strategy, Practice and Science* 1 (3): 161–168.

Phoenix C. (2003) 'Design of a Primitive Nanofactory', *Journal of Evolution and Technology* 13 (2). Online, available at: www.jetpress.org/volume13/Nanofactory.htm (accessed 7 Nov. 2003).

Pillsbury M. (ed.) (1998) *Chinese Views of Future Warfare*, rev. edn, Washington DC: National Defense University Press. Online, available at: www.ndu.edu/inss/books/books%20_%201998/Chinese%20Views%20of%20Future%20Warfare%20-Sept%2098/chinacont.html (accessed 27 Oct. 2003).

—— (2000) *China Debates the Future Security Environment*, Washington DC: National Defense University Press. Online, available at: www.ndu.edu/inss/books/pills2.htm (accessed 31 Oct. 2000).

Pomrenke G. (2002) 'US National Nanotechnology Initiative & the DoD', viewgraphs presented at *Defence Nanotechnology 2002*, 31 Oct.–1 Nov., London: Defence Event Management.

QinetiQ (2003) 'QinetiQ Nanomaterials – Ignitors'. Online, available at: www.nano.qinetiq.com/03_applications_knowledge/showcase.asp (accessed 18 Nov. 2003).

Reed M.A. and Tour J.M. (2000) 'Computing with Molecules', *Scientific American* 282 (6): 69–75.

Rees M. (2003) *Our Final Hour*, New York: Basic.

Reifman E.M. (1996) 'Diamond Teeth', in B.C. Crandall (ed.), *Nanotechnology – Molecular Speculation on Global Abundance*, 3rd printing, Cambridge MA/London: MIT Press.

Reip P. (2002) 'Nanomaterials in Defence', viewgraphs presented at *Defence Nanotechnology 2002*, 31 Oct.–1 Nov., London: Defence Event Management.

Rennie G. (2003) 'Predicting Stability for the High-Energy Buckyball', *Science & Technology Review* (December), 20–21.

Reynolds G.H. (2002) *Forward to the Future: Nanotechnology and Regulatory Policy*, San Francisco CA: Pacific Research Institute (November 2002). Online, available at: www.pacificresearch.org/pub/sab/techno/forward_to_nanotech.pdf (accessed 24 Jan. 2003).

Rissanen J. (2002) 'Left in Limbo: Review Conference Suspended On Edge of Collapse', *Disarmament Diplomacy*, no. 62 (January–February). Online, available at: www.acronym.org.uk/dd/dd62/62bwc.htm (accessed 24 Sept. 2002).

Robotics (1997) Robotics Workshop 2020, US Army Research Laboratory, Jet Propulsion Laboratory, Pasadena CA, 25–27 February 1997, McLean VA: Science Applications International.

Roco M.C. (1999) 'Research Programs on Nanotechnology in the World', in R.W. Siegel, E. Hu and M.C. Roco (eds) *Nanostructure Science and Technology – A Worldwide Study*, Baltimore MD: WTEC, Loyola College. Also: Boston etc.: Kluwer. Online, available at: www.wtec.org/loyola/pdf/nano.pdf (accessed 25 April 2003).

—— (2001) 'International Strategy for Nanotechnology Research and Development', *Journal of Nanoparticle Research* 3 (5–6): 353–360.

—— (2002) 'National Nanotechnology Investment in the FY 2003 Budget Request by the President'. Online, available at: nano.gov/2003budget.html (accessed 10 Sept. 2002).

—— (2002a) 'National Nanotechnology Initative and a Global Perspective', paper presented at 'Small Wonders' NSF Symposium, March 19, 2002. Online, available at: www.nsf.gov/home/crssprgm/nano/smwonder_slide.pdf (accessed 17 Sept. 2002).

—— (2003) 'Government Nanotechnology Funding: An International Outlook', NSF. Online, available at: www.nano.gov/intpersp_roco_june30.htm (accessed 28 Oct. 2003).

—— (2003a) 'Broader Societal Issues of Nanotechnology', *Journal of Nanoparticle Research* 5 (3–4): 181–189.

Roco M.C. and Bainbridge W.S. (eds) (2001) *Societal Implications of Nanoscience and Nanotechnology* (NSF Report), Boston etc.: Kluwer. Online, available at: www.wtec.org/loyola/nano/societalimpact/nanosi.pdf (accessed 22 Sept. 2003).

—— (2003) *Converging Technologies for Improving Human Performance – Nanotechnology, Biotechnology, Information Technology and Cognitive Science*, Boston etc.: Kluwer. Online, available at: www.wtec.org/ConvergingTechnologies/NBIC_report.pdf (accessed 29 Oct. 2003).

Roco M.C. and Tomellini R. (eds) (2002) *Nanotechnology – Revolutionary Opportunities and Societal Implications*, Luxembourg: European Communities. Parts online, available at: www.cordis.lu/nanotechnology/src/publication.htm (accessed 30 Aug. 2002).

Roman C. (2002) *It's Ours to Lose – An Analysis of EU Nanotechnology Funding and the Sixth Framework Programme*, Brussels: European Nanobusiness Association. Online, available at: www.nanoeurope.org/docs/European%20Nanotech%20Funding.pdf (accessed 10 Nov. 2003).

Rossetto O., Rigoni M. and Montecucco C. (2004) 'Different Mechanism of Blockade of Neuroexocytosis by Presynaptic Neurotoxins', *Toxicology Letters* 149 (1–3): 91–101.

Royal (2003) Royal Society, 'Nanotechnology – Request for Initial Views on this New Study'. Online, available at: www.royalsoc.ac.uk/nanotechnology/intro.htm (accessed 14 July 2003).

Rudd C.D. and Shaw R.W. (eds) (2001) 'Nanostructures in Polymer Matrices', Report on a Workshop, University of Nottingham, 10–13 Sept. 2001, London: US Army European Research Office. Online, available at: www.nano.gov/ch4_f01_polymers_workshop.doc (accessed 27 Jan. 2003).

Rudolph A. (2001) 'The Brain-Machine Interface'. Online, available at: www.darpa.mil/dso/thrust/sp/presentations/BioMagnetics_Rudolph.pdf (accessed 16 Jan. 2003).

Russell S.J. and Norvig P. (2003) *Artificial Intelligence – A Modern Approach*, 2nd edn, Upper Saddle River NJ: Pearson Education.

Russia (2000) 'National Security Concept of the Russian Federation', translation from *Rossiiskaya Gazeta*, January 18, 2000. Online, available at: www.fas.org/nuke/guide/russia/doctrine/gazeta012400.htm (accessed 5 April 2004).

—— (2000a) 'Military Doctrine of the Russian Federation', Presidential Decree of April 21, 2000. Online, available at: www.dcaf.ch/publications/e-publications/Rus_legal_acts/04_Military_doctrine.pdf (accessed 5 April 2004).

Salvetat-Delmotte J.-P. and Rubio A. (2002) 'Mechanical Properties of Carbon Nanotubes: a Fiber Digest for Beginners', *Carbon* 40 (10): 1729–1734.

Savage S.J. (2002) 'Evolution of a Swedish Defence Nanotechnology Programme', viewgraphs presented at *Defence Nanotechnology 2002*, 31 Oct.–1 Nov., London: Defence Event Management.

Scheffran J. (1985) *Rüstungskontrolle im Weltraum – Risiken und Verifikationsmöglichkeiten bei Anti-Satelliten-Waffen*, Schriftenreihe des Arbeitskreises Marburger Wissenschaftler für Friedens- und Abrüstungsforschung Nr. 1, Marburg, Fachbereich Physik der Philipps-Universität (shortened: HSFK-Report 6/1986, Frankfurt/M.: Hessische Stiftung Friedens- und Konfliktforschung).

—— (1986) 'Verification and Risk for an Antisatellite Weapons Ban', *Bulletin of Peace Proposals* 17 (2): 165–174.

Schneiker C. (1989) 'NanoTechnology With Feynman Machines: Scanning Tunneling Engineering and Artificial Life', in C.G. Langton (ed.) *Artificial Life*, Santa Fe Institute Studies in the Sciences of Complexity vol. VI, Reading MA etc.: Addison-Wesley.

Scientific (2001) *Nanotech – The Science of the Small Gets Down to Business*, Special Issue, Scientific American 285 (3) (German: *Nanotechnologie*, Spektrum der Wissenschaft Spezial 2/2001).

Service R.F. (2003) 'Next-Generation Technology Hits an Early Midlife Crisis', *Science* 302 (5645): 556–559.

Shaw R.W. (2002) *European Nanomaterials Research and the Army*, 23rd Army Science Conference, Dec. 2–5, 2002, Orlando FL. Online, available at: www.asc2002.com/summaries/I/LP-12.pdf (accessed 22 April 2003).

SIA (2001) Semiconductor Industry Association, *The International Technology Roadmap for Semiconductors*, 2001 Edition, Executive Summary, Austin TX: International SEMATECH. Online, available at: public.itrs.net/Files/2001ITRS/ExecSum.pdf (accessed 14 April 2003).

Siegel R.W., Hu E. and Roco M.C. (1999) *Nanostructure Science and Technology – A Worldwide Study*, Baltimore MD: WTEC, Loyola College. Also: Boston etc.: Kluwer. Online, available at: www.wtec.org/loyola/pdf/nano.pdf (accessed 25 April 2003).

Singer C.E. and Sands A. (2002) *Keys to Unblocking Multilateral Nuclear Arms Control*, ACDIS Occasional Paper SIN: 2, Urbana-Champaign IL: University of Illinois. Online, available at: www.acdis.uiuc.edu/homepage_docs/pubs_docs/PDF_Files/SingerSandsOPjul02.pdf (accessed 21 Nov. 2002).

Smalley R.E. (2001) 'Of Chemistry, Love and Nanobots', *Scientific American* 285 (3): 76–77.

Smith H.O., Hutchison C.A. III., Pfannkoch C. and Venter J.C. (2003) 'Generating a Synthetic Genome by Whole Genome Assembly: ϕX174 Bacteriophage from Synthetic Oligonucleotides', *Proceedings of the National Academy of Sciences USA* 100 (26): 15440–15445.

Smith R.H. (2001) 'Social, Ethical, and Legal Implications of Nanotechnology', in M.C. Roco and W.S. Bainbridge (eds) *Societal Implications of Nanoscience and Nanotechnology* (NSF Report), Boston etc.: Kluwer. Online, available at: www.wtec.org/loyola/nano/societalimpact/nanosi.pdf (accessed 22 Sept. 2003).

Smith II R.H. (1998) *A Policy Framework for Developing a National Nanotechnology Program*, MSc thesis, Virginia Polytechnic Institute and State University. Online, available at: www.altfutures_afa.com/nanotech/smithlinks/thesis.pdf (accessed 24 Jan. 2003).

SNL (2003) 'Nanoscience & Nanotechnology at Sandia National Laboratories'. Online, available at: nano.sandia.gov (accessed 27 Jan. 2003).

Snow E. (2003) 'Nanoelectronics Overview'. Online, available at: nanoscience.nrl.navy.mil/files/NanoOverview.zip, . . .ppt (accessed 15 April 2003).

Solem J.C. (1994) 'The Motility of Microrobots', in C.G. Langton (ed.) *Artificial Life III – Proceedings of the Workshop on Artificial Life held June 1992 in Santa Fe, New Mexico*, Reading MA etc.: Addison-Wesley.

Soong R.K., Bachand G.D., Neves H.P., Olkhovets A.G., Craighead H.G. and Montemagno C.D. (2000) 'Powering an Inorganic Nanodevice with a Biomolecular Motor', *Science* 290 (5496): 1555–1558.

SSHP (2001) 'Small Satellites Home Page'. Online, available at: www.ee.surrey.ac.uk/SSC/SSHP (accessed 29 June 2001).

Steinbruner J. and Lewis J. (2002) 'The Unsettled Legacy of the Cold War', *Dædalus* 131 (4): 5–10.

Steinbruner J.D. and Harris E.D. (2003) 'Controlling Dangerous Pathogens', *Issues in Science and Technology* 19 (3): 47–54.

Suchman M.C. (2001) 'Envisioning Life on the Nano-Frontier', in M.C. Roco and W.S. Bainbridge (eds) *Societal Implications of Nanoscience and Nanotechnology* (NSF Report), Boston etc.: Kluwer. Online, available at: www.wtec.org/loyola/nano/societalimpact/nanosi.pdf (accessed 22 Sept. 2003).

Sun B. (1996/1998) 'Nanotechnology Weapons on Future Battlefields', pp. 413–420 in M. Pillsbury (ed.) *Chinese Views of Future Warfare*, rev. edn, Washington DC: National Defense University Press. Online, available at: www.ndu.edu/inss/books/books%20_%201998/Chinese%20Views%20of%20Future%20Warfare%20-Sept%2098/chinapt4.html (accessed 27 Oct. 2003).

Sunshine (2002) *Non-Lethal Weapons Research in the US: Genetically Engineered Anti-Material Weapons*, Backgrounder Series #9, Sunshine Project, March 2002. Online, available at: www.sunshine-project.org/publications/bk/bk9en.html (accessed 5 April 2004).

—— (2002a) 'US Special Forces Seek Genetically Engineered Bioweapon', News Release, Sunshine Project, 12 Aug. 2002. Online, available at: www.sunshine-project.org/publications/pr/pr120802.html (accessed 5 April 2004).

—— (2003) *Biosafety, Biosecurity, and Bioweapons – Three Agreements on Biotechnology, Health, and the Environment, and Their Potential Contribution to Biological Weapons Control*, Background Paper #11, The Sunshine Project (October 2003). Online, available at: www.sunshine-project.org/publications/bk/bk11.html (accessed 27 Oct. 2003).

—— (2004) 'The Return of the ARCAD', News Release, Sunshine Project. Online, available at: www.sunshine-project.org/publications/pr/pr060104.html (accessed 9 Feb. 2004).

—— (2004a) 'Milzbrand- und Mehrfachimpfungen von SoldatInnen zunehmend unter Beschuss', *Biowaffen-Telegramm* Nr. 24, Sunshine-Project (16 Jan. 2004).

Talbot D. (2002) 'Supersoldiers', *Technology Review* (October): 44–51.

Talwar S.K., Xu S., Hawley E.S., Weiss S.A., Moxon K.A. and Chapin J.K. (2002) 'Rat Navigation by Remote Control', *Nature* 417 (6884): 37–38.

Taniguchi N. (ed.) (1996) *Nanotechnology – Integrated Processing Systems for Ultra-precision and Ultra-fine Products*, Oxford etc.: Oxford University Press.

Tenner E. (2001) 'Nanotechnology and Unintended Consequences', in M.C. Roco and W.S. Bainbridge (eds) *Societal Implications of Nanoscience and Nanotechnology* (NSF Report), Boston etc.: Kluwer. Online, available at: www.wtec.org/loyola/nano/societalimpact/nanosi.pdf (accessed 22 Sept. 2003).

Timbrell J.A. (1989) *Introduction to Toxicology*, London etc.: Taylor & Francis.

Timp G. (ed.) (1999) *Nanotechnology*, New York: AIP/Springer.

Timp G., Howard R.E. and Mankiewich P.M. (1999) 'Nano-electronics for Advanced Computation and Communication', in G. Timp (ed.) *Nanotechnology*, New York: AIP/Springer.

Tolles W.M. (2001) 'National Security Aspects of Nanotechnology', in M.C. Roco and W.S. Bainbridge (eds) *Societal Implications of Nanoscience and*

Nanotechnology (NSF Report), Boston etc.: Kluwer. Online, available at: www.wtec.org/loyola/nano/societalimpact/nanosi.pdf (accessed 22 Sept. 2003).

Tucker J.B. (2003) 'Preventing the Misuse of Pathogens: The Need for Global Biosecurity Standards', *Arms Control Today* 33 (5): 3–10.

Tyler P.E. (1992) 'U.S. Strategy Plan Calls For Insuring No Rivals Develop', *New York Times*, 8 March 1992.

UCR (2002) University of California, Riverside, News Release, 'New center for nanoscale innovation transfers knowledge from universities to industry' (Dec. 10, 2002). Online, available at: www.newsroom.ucr.edu/cgi-bin/display.cgi?id=305 (accessed 3 Feb. 2003).

UK MoD (2001) *The Future Strategic Context for Defence*, London: Ministry of Defence. Online, available at: www.mod.uk/index.php3?page=2449 (accessed 13 Feb. 2001).

—— (2001a) *Nanotechnology: Its Impact on Defence and the MoD*, Nov. 2001. Online, available at: www.mod.uk/linked_files/nanotech.pdf, see also http://www.mod.uk/issues/nanotech/contents.htm (accessed 18 Nov. 2003).

Ullmanns (1982) *Ullmanns Encyklopädie der technischen Chemie* (4th edn), Weinheim etc.: Verlag Chemie (vol. 21, Chapter 'Sprengstoffe').

Vettiger P., Cross G., Despont M., Drechsler U., Dürig U., Gotsmann B., Häberle W., Lantz M.A., Rothuizen H.E., Stutz R. and Binnig G.K. (2002) 'The "Millipede" – Nanotechnology Entering Data Storage', *IEEE Transactions on Nanotechnology* 1 (1): 39–55.

Vinge V. (1993) 'The Singularity', paper originally presented at VISION-21 Symposium, March 30–31, 1993, slightly changed in *Whole Earth Review*, Winter 1993. Online, available at: www.ugs.caltech.edu/~phoenix/vinge/vinge-sing.html (accessed 28 April 2003).

Vogel V. (2001) 'Societal Impacts of Nanotechnology in Education and Medicine', in M.C. Roco and W.S. Bainbridge (eds) *Societal Implications of Nanoscience and Nanotechnology* (NSF Report), Boston etc.: Kluwer. Online, available at: www.wtec.org/loyola/nano/societalimpact/nanosi.pdf (accessed 22 Sept. 2003).

Wang J. and Dortmans P.J. (2004) *A Review of Selected Nanotechnology Topics and Their Potential Military Applications*, DSTO-TN-0537, Edinburgh South Australia: Systems Sciences Laboratory, Defence Science and Technology Organisation (Jan.). Online, available at: www.dsto.defence.gov.au/corporate/reports/DSTO-TN-0537.pdf (accessed 30 April 2004).

Weil V. (2001) 'Ethical Issues in Nanotechnology', in M.C. Roco and W.S. Bainbridge (eds) *Societal Implications of Nanoscience and Nanotechnology* (NSF Report), Boston etc.: Kluwer. Online, available at: www.wtec.org/loyola/nano/societalimpact/nanosi.pdf (accessed 22 Sept. 2003).

Wessberg J., Stambaugh C.R., Kralik J.D., Beck P.D., Laubach M., Chapin J.K., Kim J., Biggs S.J., Srinivasan M.A. and Nicolelis M.A.L. (2000) 'Real-time Prediction of Hand Trajectory by Ensembles of Cortical Neurons in Primates', *Nature* 408 (6810): 361–365.

Wheelis M. (2003) '"Nonlethal" Chemical Weapons: A Faustian Bargain', *Issues in Science and Technology online* 19 (3). Online, available at: www.issues.org/19.3/wheelis.htm (accessed 30 April 2004).

Whitesides G.M. (2001) 'The Once and Future Nanomachine', *Scientific American* 285 (3): 78–83.

Whitesides G.M. and Love J.C. (2001) 'Implications of Nanoscience for Knowledge and Understanding', in M.C. Roco and W.S. Bainbridge (eds) *Societal Implications of Nanoscience and Nanotechnology* (NSF Report), Boston etc.: Kluwer. Online, available at: www.wtec.org/loyola/nano/societalimpact/nanosi.pdf (accessed 22 Sept. 2003).

Williams R.S. (2002) 'How can the Intersection of Nano and Information Technologies Change the Society?' in M.C. Roco and R. Tomellini (eds) *Nanotechnology – Revolutionary Opportunities and Societal Implications*, Luxembourg: European Communities. Parts online, available at: www.cordis.lu/nanotechnology/src/publication.htm (accessed 30 Aug. 2002).

Williams R.S. and Kuekes P.J. (2001) 'We've only just Begun', in M.C. Roco and W.S. Bainbridge (eds) *Societal Implications of Nanoscience and Nanotechnology* (NSF Report), Boston etc.: Kluwer. Online, available at: www.wtec.org/loyola/nano/societalimpact/nanosi.pdf (accessed 22 Sept. 2003).

Wilson T. (2001) 'Threats to United States Space Capabilities', Prepared for the Commission to Assess United States National Security Space Management and Organization (Appendix to Report of Jan. 11, 2001). Online, available at: www.space.gov/commission/support-docs/article05.pdf (accessed 22 Jan. 2002).

Yonas G. and Picraux S.T. (2001) 'National Needs Drivers for Nanotechnology', in M.C. Roco and W.S. Bainbridge (eds) *Societal Implications of Nanoscience and Nanotechnology* (NSF Report), Boston etc.: Kluwer. Online, available at: www.wtec.org/loyola/nano/societalimpact/nanosi.pdf (accessed 22 Sept. 2003).

Zimmerman P.D. and Dorn D.W. (2002) 'Computer Simulation and the Comprehensive Test Ban Treaty', Defense Horizons, no. 17. Online, available at: www.ndu.edu/inss/DefHor/DH17/DH17.pdf (accessed 13 Dec. 2002).

Zyvex (2003) 'About Us ... Background'. Online, available at: www.zyvex.com/AboutUs/background.html (accessed 4 Nov. 2003).

—— (2003a) 'Products...'. Online, available at: www.zyvex.com/Products/home.html (accessed 4 Nov. 2003).

INDEX